PERGAMON INSTITUTE OF E

Materials for Language Practice

BIOLOGICAL SCIENCES

Developing Reading Skills in English

Other titles in this series include

DAVIES, D.
Telecommunications

JOYCEY, E.
Docks and Harbours

NOTO, M.
Physics

PRODROMOU, L.
Medicine

Other ESP titles of interest include

BINHAM, P. *et al.*
*Hotel English**

BINHAM, P. *et al.*
*Restaurant English**

BRIEGER *et al.*
*Business Contacts**

BRIEGER, N. and J. Comfort
Business Issues

DUNLOP, I. and H. Schrand
Communication for Business

FITZPATRICK, A.
*English for International Conferences**

GUDERIAN, C.
*Hold On!**

LEBAUER, R.
Reading Skills for the Future

McGOVERN, J. and J. McGovern
*Bank on Your English**

McKELLEN, J. and M. Spooner
*Business Matters**

POTE *et al.*
*A Case for Business English**

PALSTRA, R.
*International Telephone Users' Handbook**

*includes audio cassette(s)

BIOLOGICAL SCIENCES

Developing Reading Skills in English

Valerie Kay

PERGAMON INSTITUTE OF ENGLISH

a member of the Pergamon Group

Oxford · New York · Toronto · Sydney · Paris · Frankfurt

U.K.	Pergamon Press Ltd., Headington Hill Hall, Oxford OX3 0BW, England
U.S.A.	Pergamon Press Inc., Maxwell House, Fairview Park, Elmsford, New York 10523, U.S.A.
CANADA	Pergamon Press Canada Ltd., Suite 104, 150 Consumers Road, Willowdale, Ontario M2J 1P9, Canada
AUSTRALIA	Pergamon Press (Aust.) Pty. Ltd., P.O. Box 544, Potts Point, N.S.W. 2011, Australia
FRANCE	Pergamon Press SARL, 24 rue des Ecoles, 75240 Paris, Cedex 05, France
FEDERAL REPUBLIC OF GERMANY	Pergamon Press GmbH, Hammerweg 6, D-6242 Kronberg Taunus, Federal Republic of Germany

First edition 1985

Library of Congress Cataloging in Publication Data

Kay, Valerie.
Biological sciences.
(Materials for language practice)
1. Readers—Biology. 2. English language—Text-books for foreign speakers. 3. Biology—Addresses, essays, lectures. I. Title. II. Series.
PE1127.B54K38 1984 428.6'4'024574 84–6354

British Library Cataloguing in Publication Data

Kay, Valerie
Biological sciences.—(Materials for language practice)
1. English language—Text-books for foreign speakers. 2. Readers—1950–
I. Title II. Series
428.2'4 PE1128

ISBN 0–08–031063–X

Printed in Great Britain by Redwood Burn Ltd., Trowbridge

Contents

Introduction

This book, which is one of a series, is intended for students of biological sciences whose mother tongue is not English but who need an adequate reading performance in English as a tool for advancing their studies in their own particular discipline.

The texts

The texts have been chosen in order to offer the student a broad range of language within the limits of the subject, and at different levels of difficulty. The selected passages range from extracts taken from relatively elementary textbooks, such as would be used in the secondary school system, through standard works at University level, by way of some "popular" science as a little light relief, to advanced papers published in learned journals. The topics are representative of biological sciences in the broadest sense, and include texts relating to plant and animal physiology, genetics, environmental studies, applied biology, microbiology, embryology and virology. Each text is entirely authentic; that is to say, nothing has been "simplified" in a misguided effort to make it more accessible to the foreign reader. What the student is being offered is exactly what the native speaker of English would encounter.

The purpose of the book is to build up the readers' confidence, with a variety of tasks, so that when, in the course of their studies, they face an unfamiliar text in English, they will have at their disposal techniques which will permit them access to the information contained in the text in *the least possible time.* The intention is not only to help the students to read more fluently in English, but also to improve their overall reading speed.

The exercises

All the exercises in this book have been designed to fulfil a specific purpose in training the student to acquire or develop the skills and techniques necessary to become a fast and fluent reader—one who is able to determine quickly what needs to be extracted from a given text, who can assess to what extent a general or "global" understanding will be enough and when (and how far) it is essential to go into detail. The exercises fall broadly into three categories:

1. Those designed to help the student to read better, and by improving reading techniques, to read faster.
2. Those designed to help the reader to understand the text and to deduce the meanings of unfamiliar words.
3. Production exercises, sometimes requiring written work.

Exercise types

In Category 1, there are the types which involve the reading techniques of *skimming* and *scanning*. In both cases, the reader must go through the text very quickly, not pausing over every unknown word. In skimming, however, the idea is to get a general grasp, in the broadest terms, of what a text is about. In scanning, the reader is searching for specific details. Other exercises in this category require the reader to make educated guesses as to what has gone before, to predict what comes next, to suggest a probable continuation of a text. There are also information transfer exercises, and exercises on the finding of key points.

In Category 2, there are multiple choice questions, "fill the blank" (cloze) tests and matching exercises to help the student appreciate sentence structure, aids to understanding of unfamiliar words, and a number of exercises in helping the student to understand references within a passage and the logical organization of a text.

In Category 3, the emphasis is on the guided construction of paragraphs and the writing of summaries.

How to use the book

Before starting to read any of the texts, the students should first look at the accompanying exercises to see what will be required from their reading of the passage. They should look particularly at the first exercise, which, in most Units, will be a skimming or scanning exercise. It is advisable to work through the exercises in order, as, in general, they are arranged to lead from a "global" view, through a detailed examination to a reformulation of the main points. When the exercises for a Unit have been completed, and the answers checked against the Key (see below), the text can be used later in a controlled procedure designed to increase reading speed. Once the Unit has been exhaustively studied, the text should be put aside for several days. Then the student should read it again, straight through, making a careful note of the time taken. A log of the results should be kept, showing the length of the passage (number of words) and the reading speed in minutes and seconds. The student can then plot progress with each succeeding text. Two things should be stressed at this point:

1. In work of this nature, the student is in competition with himself, or at least against the clock, and is in no way competing with other members of a group or class.
2. The actual reading should always be silent; reading aloud by a student should play no part in any classroom activity aimed at achieving a reading competence. It is a highly specialized skill, and not one that is frequently called for in ordinary life. It may occasionally be necessary for a teacher to read a text to a group, but in the main, for the student reading is a private, personal and, above all, silent proceeding.

Classwork

Where the book is being used with a class, the teacher should encourage the students to work in pairs, or small groups, depending on the type of exercise. The skimming or scanning exercises could be tackled by two students working together, while most of the rest could very well be done by group work. A small group of, say, four students can exchange ideas and reach a consensus. Then each group's answer could be examined by the other groups.

The student alone

The materials are also suitable for the student working alone, and can be used for unsupervised self-study. In such a case it is imperative that the student uses the Answer Key in a constructive way (see below).

The answer key

A set of answers to the exercises in provided, and it is vital that proper attention be paid to the checking of answers against this key. Where there is no one "right" answer, several possible ones are suggested. In all cases, however, the student must appreciate why one answer is to be preferred over another, and why an answer can be rejected as wrong.

The answers must be studied, especially by the student working alone. It is not enough just to accept that a question may have been wrongly answered, and then pass on. The student should return to the text, examine it again until the reason for the right answer becomes apparent. In class, correcting exercises should be done carefully, with adequate time for discussion, and perhaps protest and dissension. It must not be a mere "rubber-stamping" of ticks and crosses.

Acknowledgements

Acknowledgements are given after each text. The publishers are grateful for the kind permissions given to reproduce extracts.

Every effort has been made to trace and acknowledge ownership of copyright and the publishers will be glad to make suitable arrangements with any copyright holders whom it has not been possible to contact.

UNIT 1

Respiration

The modern definition of respiration is: the processes which lead to, and include, the chemical breakdown of materials to provide energy for life. These processes occur inside the living cells of every type of organism.

To avoid unnecessary confusion, it is strongly recommended that respiration be used only in this modern sense, so that it is clearly distinguished from the mechanisms of breathing, which is concerned with the absorption of oxygen from the air.

The release of energy

The energy for life is released during respiration from substances known loosely as "food". There are many different foods, and they are taken into the body in many different ways, but in the majority of organisms all foods are converted into glucose sugar before they are used as a source of energy. For the sake of simplicity, the following descriptions refer to the respiration of glucose.

In most organisms energy is released by a process called aerobic respiration, which requires a continuous supply of oxygen molecules obtained from the air or water surrounding the organism. In certain circumstances, however, energy can be released without the use of oxygen molecules. This is known as anaerobic respiration. These two different but related types of respiration are described in the following sections.

Aerobic respiration

The aerobic respiration of glucose is summarized by the following chemical equation:

$$C_6H_{12}O_6 + 6O_2 \text{ —— } 6CO_2 + 6H_2O + 2898\text{kJ of energy glucose}$$

Aerobic respiration releases all the available energy within each glucose molecule; that is, it produces the same amount of energy that is released when glucose is burnt in oxygen gas. The chemical equation above gives the false impression that respiration involves only one chemical reaction, because it shows only the raw materials and end-products of respiration. The whole process involves a sequence of some fifty separate reactions, each catalysed by a different enzyme. The result is a controlled release of energy which is far more useful to the organism than a sudden explosive burst of energy.

Look again at the equation, and see what happens to all the hydrogen atoms contained within a glucose molecule. Eventually these atoms combine with oxygen atoms to form water. In fact, the bulk of respiratory energy becomes available to the organism as hydrogen atoms are removed from glucose during respiration. This process is catalysed mainly by dehydrogenase enzymes. In other words, the oxygen which an aerobic organism has absorbed combines with hydrogen atoms from glucose or other foods to produce water, which may be excreted from the body.

Anaerobic respiration

Anaerobic respiration differs from aerobic respiration in three important ways. First, anaerobic reactions break down glucose in the absence of oxygen. Second, anaerobic reactions do not completely break down glucose into carbon dioxide and water but into intermediate substances such as lactic acid or alcohol. Third, anaerobic respiration releases far less energy than aerobic respiration, because glucose is not completely broken down.

Organisms which respire anaerobically are called anaerobes. Certain bacteria are complete anaerobes. They live permanently in conditions where no oxygen exists and rely entirely upon anaerobic respiration for energy.

[Extract from B. S. Beckett, *Biology, A Modern Introduction*, OUP 1976. By permission of Oxford University Press.]

EXERCISES

1.1 Skimming

Read through the passage very quickly. Do not try to understand every word. The intention is to get a general idea of what the passage is about. Look at the title of the book from which this extract has been taken. Now choose what you think are the best answers to the following questions:

(a) What type of reader was the original book intended for?
  (i) students ✓
  (ii) teachers

(b) How much does the author expect the reader to know about respiration?
   (i) nothing
   (ii) a lot
   (iii) some basic information
(c) How do you think the original text continued?
   (i) with more information about anaerobic respiration
   (ii) by returning immediately to the subject of aerobic respiration.

1.2 Scanning

This technique also requires you to read through the passage quickly, but this time looking for particular details and specific points. Find the answers to the following questions.

(a) What process is concerned with the absorption of oxygen from the air?
(b) What totally anaerobic organisms are mentioned?
(c) What are the principal catalysts in the process of producing water from a glucose molecule?
(d) What type of respiration occurs in the majority of organisms?
(e) Is the description of respiration given in the text applicable to all "foods"?
(f) How much available energy does aerobic respiration release from a glucose molecule?
(g) In how many important ways are aerobic and anaerobic respiration different?
(h) Does aerobic respiration release more or less energy from glucose than anaerobic respiration?

1.3 Understanding the text: True or False

In this type of exercise, you are given a number of statements. You have to decide which of these statements are correct, according to the text, and which are not correct.

(a) An explosive burst of energy is less useful to an organism than a controlled release.
(b) Most organisms must have a constant supply of oxygen for the release of energy.
(c) All bacteria are complete anaerobes.
(d) The original meaning of the word "respiration" was the process of breathing.
(e) All food is absorbed in the same way.
(f) The major source of energy is glucose.
(g) Only one chemical reaction is involved in respiration.

(h) Some 50 enzymes act as catalysts in the process of respiration.
(i) Water is a by-product of respiration.
(j) Anaerobic respiration is less efficient than aerobic in releasing energy from glucose.

1.4 Understanding the text: Synonyms

Find words in the text which have similar meanings to the words listed below, and which could replace them in the text. The first one has been done for you as an example, but there is another word *for it in the text, too.*

meaning = definition
happen ....
misunderstanding ....
changed ....
liberated ....
needs ....
inside ....
joins ....
totally ....
depend ....

1.5 Paragraph structure: Punctuation

All the punctuation has been removed from this passage. Put in all the necessary punctuation marks, not forgetting capital letters where required.

when flooding occurs for example plants can survive for several days completely immersed in water they can survive for several weeks in waterlogged airless soil anaerobic respiration is also necessary in the initial stages of germination when the plant embryo is completely enclosed within an airtight seedcoat the muscles of vertebrate animals can continue working for a minute or two without oxygen this happens for example when an athlete runs in a race for the first few yards his muscles respire aerobically but soon they use up all their available oxygen

1.6 Writing a summary

Fill in the blanks to make a short summary of the information contained in the text.

Respiration is the chemical _______ of materials to provide _______ for _______ . Aerobic respiration requires a _______ of _______ and results in the _______ breakdown of _______ into carbon dioxide and _______ with the release of _______ . Anaerobic _______ requires no _______ , produces far less _______ , and breaks down _______ into intermediate _______ such as _______ and _______ .

UNIT 2

DNA and the Genetic Code

Genes are the instructions which direct the growth and development of organisms. The popular name for these instructions is the genetic code. In 1944 it was shown that genes are made of a chemical called deoxyribonucleic acid or DNA. In the early 1950s the structure of DNA was discovered and this led to an understanding of the genetic code.

Structure of DNA

DNA is a long, thread-like molecule similar in shape to a rope ladder twisted into a spiral. The upright sides of the ladder are made of alternate sugar and phosphate molecules joined into a chain. The rungs of the ladder are made of chemical bases, of which there are four types: adenine, guanine, cytosine, and thymine. For simplicity these bases are usually referred to by their initial letters A, G, C, and T. (See illustration for Exercise 5.)

The genetic code

The genetic code is formed by the sequence in which these four chemical bases are arranged along the length of a DNA molecule. Put another way, if the genetic code were a set of written rather than chemical instructions, the letters A, G, C, and T would be the alphabet in which the "words" of the code were written. In fact, the code is written in words of only three letters each; that is, it consists of groups of three adjacent bases. For example, the base sequence AAT along one strand is one word, CAA is another, and so forth. The number of different three-letter words which can be composed out of the four letters A, G, C, and T is 64.

It is now known that these three-letter words (groups of three bases) are the instructions to a cell telling it how to make protein molecules out of amino acids. The sequence AAT (adenine, adenine, thymine) in the code tells a cell how to make the amino acid leucine, and CAA (cytosine, adenine, adenine) tells a cell how to make the amino acid valine.

A single gene consists of a thousand or more groups of three bases. Together these groups represent instructions to make one complete protein molecule of a specific type. A cell "obeys" these instructions by putting together the correct amino acids in the exact sequence directed by the gene.

DNA, therefore, controls the type of protein manufactured in cells. In this way DNA controls the structure of an organism because proteins are the building materials out of which cells, tissues, and organs are made. But equally important is the fact that DNA directs the chemical activities which take place in an organism, because some proteins are enzymes and enzymes control the speed and type of chemical reactions in cells.

Duplication of DNA during cell division

Cell division by mitosis produces daughter cells with exact copies of the chromosomes present in the original cell. Before mitosis can occur in a cell the DNA molecules in its chromosomes must be duplicated exactly. Eventually these duplicates form a second set of chromosomes.

[Extract from B.S. Beckett, *Illustrated Human and Social Biology*, OUP 1981. By permission of Oxford University Press.]

EXERCISES

2.1. Skimming

Read through the passage quickly. Try to get a general idea of what it is about. Now choose a new title for the passage from the following suggestion:

(a) The language of chemistry
(b) Instructions for growth
(c) A chemical language
(d) How cells obey orders
(e) Learning a new alphabet

2.2 Scanning

Read through the text quickly, but this time look for details and answer the following questions:

(a) When was the structure of DNA discovered?
(b) To which chemical bases do the letters G and T refer?
(c) What are the sides of the DNA "ladder" made of?
(d) What instructions are given to a cell by "three-letter words"?

(e) What controls the speed and type of chemical reactions in cells?
(f) What process produces daughter cells with exact copies of the chromosomes present in the original cell?

2.3 Understanding the text: True or False?

Decide which of these statements are correct and which are not correct, according to the information contained in the passage.

(a) Genes control the size of organisms.
(b) The genetic code was first understood in 1944.
(c) The rungs of a ladder join its two sides together.
(d) The letters of the genetic code can form 64 four-letter words.
(e) The instruction to form amino acid leucine is AAT.
(f) Proteins are the materials from which organisms are made.

2.4 Sentence structure

This type of exercise helps you to understand how sentences are constructed and to appreciate which class of word is required in each case.

Before describing DNA duplication ______ is necessary to look ______ closely at its structure. ______ rungs of the ladder-like ______ molecule are made up ______ chemical bases represented by A, G, C ______ T. Each rung consists of ______ bases joined at the ______ of the ladder. There ______ millions of rungs in ______ DNA molecule, but in ______ one, base A is ALWAYS ______ base T and base C ______ ALWAYS opposite base G. This ______ known as the pairing ______ .

2.5 Using information

Complete the labels on this diagram. Replace the numbers with the correct word or letter.

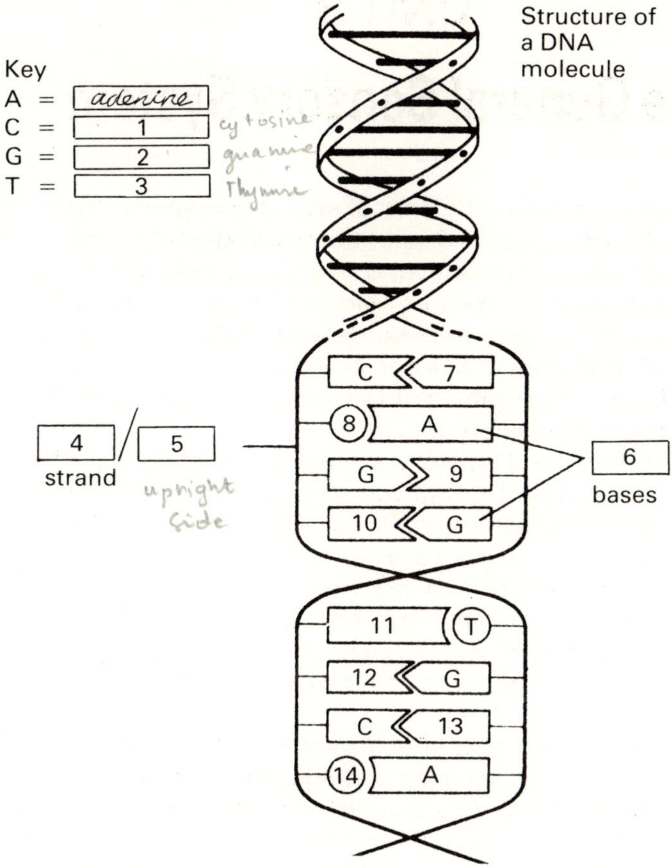

2.6 Sentence structure

Here are the elements of 5 sentences. Put them in the right order to form these sentences, then re-arrange the sentences to construct a short paragraph.

(a) on a DNA molecule/for making/each gene/a particular type of protein/carries instructions

(b) this substance/two types/of/there are

(c) by/this task/ribonucleic acid/is carried out

(d) messenger RNA/transfer RNA/they are called/and

(e) from the nucleus/the coded instructions/to the cytoplasm/for protein manufacture/pass

UNIT 3

The General Sensory System

The sensory system makes the animal aware, though not necessarily in the sense of "conscious", of conditions and changes both outside and inside its body. In the simpler animals only very general stimuli such as light or darkness, heat or cold can be perceived by the sensory system. In the higher animals detailed information about surroundings such as distance, size and colours of objects can be gained as a result of the specialization of the sensory organs and the elaboration of the nervous system.

Included in the general sensory system are the organs which are fairly evenly distributed through the dermis of the skin; hence any part of the skin is sensitive to touch, heat, pain and pressure. Examples of such sense organs are given in Fig. 1.

It must be emphasized that, in general, a particular sense organ or sense cell can respond to only one kind of stimulus. Thus, a sense organ sensitive to touch will not be affected by the stimulus of heat; a cell which is sensitive to chemicals will not respond to pressure.

It is not yet certain just how specific some of the sense organs are in their responses, for the sensory-endings which produce the sensation of pain can be activated by a variety of stimuli such as pressure, heat and cold, and the ear lobe which contains only hair plexuses and free nerve endings can detect touch, heat, cold and pressure.

Certain regions of the skin have a greater concentration of sense organs than others. The finger tips, for example, have a large number of touch organs, making them particularly sensitive to touch. The front of the upper arm is sensitive to heat and cold. Some areas of the skin have relatively few sense organs and can be pricked or burned in certain places without any sensation being felt.

Stimulation and conduction of impulses

The sense organ or sense cell is connected to the brain or spinal chord by nerve fibres. When the sense organ receives an appropriate stimulus, it sets off an electrical impulse which travels along the nerve fibre to the brain or spinal chord. When the impulse reaches one of these centres it may produce an automatic or reflex action, or record an impression by which the animal feels the nature of the stimulus and where it was applied.

The sense organs of one kind and in a definite area are connected with one particular region of the brain. It is the region of the brain to which the impulse comes that gives rise to the knowledge about the nature of the stimulus and where it was received. For example, if the regions of the brain receiving impulses from the right leg were eliminated or suppressed by drugs, no amount of stimulation of the sensory-endings would produce any sensation at all, although the sense organs would still be functioning normally. On the other hand, if a region of the brain dealing with impulses from sense organs in the leg is stimulated by any means, the sensations produced seem to be from the leg.

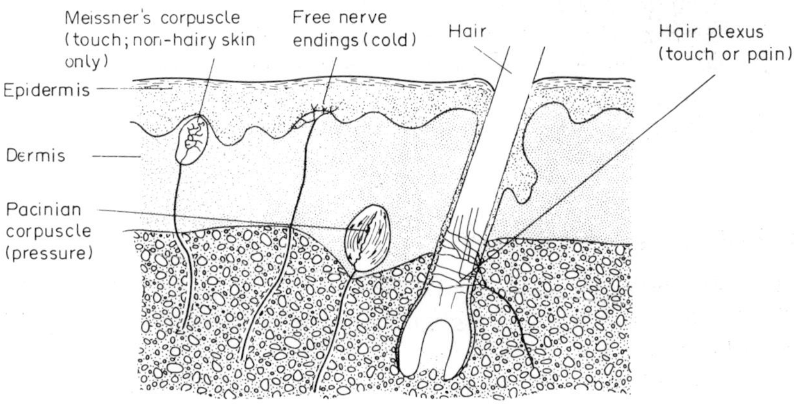

Fig. 1 Sense organs of the skin (generalized diagram)

[From *Introduction to Biology* (1982) by D. G. Mackean, John Murray Ltd.]

EXERCISES

3.1 Skimming

Read through the passage quickly and answer these questions.

(a) What type of readers was the original text intended for?

(i) teachers

(ii) students

(b) How much does the author expect the reader to know about the subject?
   (i) a lot
   (ii) very little
(c) What is an obvious clue to the level of understanding expected?

3.2 Scanning

Read through the text again, and pick out these details.

(a) What word is applied to animals capable of perceiving colour?
(b) What type of sense organs are concentrated in the finger tips?
(c) What travels along the nerve fibre after a sense organ receives the appropriate stimulus?
(d) What stimuli are said to produce the sensation of pain?
(e) Which part of the body is mentioned as being sensitive to heat and cold?
(f) Which part of the ear is sensitive to four stimuli?

3.3 Paragraph structure

Put these sentences in the right order so as to form a paragraph, and then decide whether this paragraph would have preceded or followed the given text.

(a) It is not the sensations themselves that are carried but simply a surge of electricity.
(b) Another important consideration is the fact that impulses transmitted along the nerve fibres are fundamentally all alike.
(c) For example, if the nerves from the arm and leg were changed over just before they entered the brain, stubbing the toe would feel like a pain in the arm or hand.
(d) It is only in the brain that the stimulus is identified, according to the region of the brain which the impulse enters.
(e) This is so whether it is a heat organ or a touch organ that sets off the impulse.

3.4 Understanding the text

It is often possible to understand new or unfamiliar words by recognizing words they are compared or contrasted with. For example, in the present text, if you do not know the word "aware", in the first line, you can see that it is compared with "conscious" and so it must have a similar meaning. Decide what these words are compared or contrasted with.

	COMPARED	CONTRASTED
aware	conscious	
outside		
simpler		
darkness		
heat		
burned		
automatic		
eliminated		

3.5 Understanding the text: True or False?

Decide which of these statements are correct, according to the text, and which are not correct.

(a) All parts of the skin are equally sensitive.
(b) Only the higher animals have a sensory system.
(c) Pressure-sensitive cells do not respond to chemicals.
(d) Nerve fibres connect sense cells to the brain.
(e) The stimulus to a sense cell reaches the brain as an electrical impulse.
(f) To eliminate sensation, the sense cell itself must be suppressed.

3.6 Improve your reading speed: Pointers

It is important to recognize quickly the words and phrases which are used frequently and to understand their value in the logical organization of a written text. When you see the words "for example" you know immediately that what comes next is an illustration of something that has just been mentioned. There are many of these relationships in a text. In this exercise you must identify these relationships from the "pointer" words which introduce them:

result – something happened because something else happened first

contrast – introducing an alternative or opposite

illustration – introducing an example of what has been said in order to support the point being made

logical deduction or generalization – from what has been said before

Now complete this table by putting a tick (√) in the appropriate box, to show what relationship is being indicated by the pointer words.

	Pointer	Result	Illustration	Contrast	Logical generalization Logical deduction
(a) Line 1	though				
(b) Line 6	such as				
(c) Line 10	hence				
(d) Line 13	in general				
(e) Line 14	Thus				
(f) Line 23	for example				
(g) Line 41	although				
(h) Line 42	On the other hand				

UNIT 4

Focus on Bats

Bats are furry, intelligent, social, flying mammals, spending hours grooming daily. Their wings are skin supported by arms and elongated fingers. They do not breed until their second to fourth year and produce a single baby, but not necessarily every year. Mating occurs in autumn or winter but with delayed fertilization the young are not born until June or July. The baby is raised with great care by its mother and is suckled frequently both by day and by night. At 3 weeks the young bat can fly and after 5 weeks is weaned. Bats can live for up to 30 years.

Unlike many other mammals, bats do not have a steady body temperature. In flight their body temperature of 42°C and pulse rate of 1,000 per minute are much higher than man's 37°C and 75 per minute. After landing, their temperature rapidly falls 10 degrees for digestion and later falls to the surrounding temperature. British bats feed upon insects caught in flight or picked off water, ground or foliage. During summer they consume vast numbers, many of whom are crop pests. One pipistrelle may eat over 3,500 insects each night. Because few insects are available in winter, bats put on about one-third extra weight during autumn and then hibernate from October to April with their body temperature down to zero centigrade. In Britain, bats may occasionally be seen flying in winter, often in daylight, but when they do this they may lose more weight than they can replace by feeding. When disturbed in hibernation, bats use up vital energy reserves and may die.

Bats can be found in almost any well-protected situation, but there are three major types of roost. Buildings – houses, farms, ancient monuments, churches and industrial buildings – are the most important in summer. Cave-like places – natural caves, mines, cellars, tunnels of all kinds (railway, canal and service tunnels) – are used mostly for hibernation. Hollow trees – in hedges and woodlands – are used throughout the year. Each bat species – and there are 15 resident in Britain – has its own preferred types of roost. Some are confined almost exclusively to hollow trees, while others have become adapted to using various parts of buildings such as hollow walls and roof spaces. Whatever the situation there is always at least one exit hole which may be many metres from where the bats actually roost during the day. Individual roosts are not usually occupied throughout the year, since bat colonies frequently move, but the same site tends to be occupied by the same colony at the same season each year. Distances between roosts may be only a few metres or many kilometres. The largest colonies (which rarely exceed 500 bats) are found during June to August when pregnant females all congregate in one favoured nursery roost to have their young. The bats often gather from several hundred square kilometres and therefore these roosts are vital. When weaning is complete, the adult females leave, followed later by the young. Adults begin leaving the nursery in early August and in almost every case they have gone by the end of that month. Nursery roosts are mostly on the south side of buildings.

Bats prefer clean and fairly draught-free buildings, not liking dust or cobwebs. As a result 47 per cent of colonies are found in houses less than 25 years old. For these reasons bats tend not to like old barns, although some barns do house large colonies. Roosts are selected very critically, depending on season, temperature and body conditions.

[Reproduced from *Focus on Bats, their Conservation and the Law*, courtesy of Nature Conservancy Council]

EXERCISES

4.1 Skimming

Read through the passage quickly, and try to get a general idea of what it is about. Choose from the list given below the sentence which, in your opinion, covers the main topic of the passage.

(a) Bats are dangerous to man.
(b) Old buildings are frequently damaged by bats.
(c) The text describes the life of the bat.
(d) Bats are in danger of extinction.
(e) Crop pests like bats must be killed.
(f) Animals like bats need protection.

4.2 Scanning

Now you can start to look at the text in more detail. Find words in the text which are similar in meaning to these phrases:

(a) fed on milk produced by its mother
(b) different from
(c) able to feed on normal adult food instead of mother's milk

(d) a place for very young children
(e) creatures which do damage
(f) sleeping through the winter
(g) a place where flying creatures sleep
(h) not solid

4.3 Understanding the text: True or False?

According to what is contained in the text, decide which of these statements are true, and which are false.

(a) A young bat feeds on its mother's milk for only three weeks.
(b) A bat's body temperature is never lower than a man's.
(c) British bats eat nothing but insects.
(d) Bats put on extra weight in the winter.
(e) In Britain you might see a bat flying in winter.
(f) Disturbing a hibernating bat might kill it.
(g) All bats like the same kind of place to sleep in.
(h) Colonies of bats often change from one home to another.

4.4 Understanding the text

Answer these questions either "Yes" or "No". Base your answers on information contained in the text.

(a) Do bats digest their food while they are flying?
(b) Can bats be helpful to farmers?
(c) Can bats ever find anything to eat in the winter?
(d) Are underground roosts preferred for winter sleep?
(e) Do colonies of bats usually stay in the same place for a year?
(f) Do young bats leave the roost before their parents?
(g) Do bats ever live in farm buildings?
(h) Do bats mate in the spring?

4.5 Sentence structure

Find the best word to fill in each blank.

Man has been trying to control ______ for thousands of years. These enemies ______ often both very small and very ______ . To use the term "pest" is to ______ a judgement, because one is deciding ______ a thing is good or ______ . A pest is a living organism ______ causes some kind of damage to ______ or his property. A pest is ______ which is not wanted. The problem ______ pest control is to find a ______ way of reducing the damage.

4.6 Using information

Extract information from the text to complete the following:

(a) *body temperature*: What would be a typical British bat's temperature, at an ambient temperature of 18°C?

in flight	°C
on landing	°C
after digestion	°C
during hibernation (October – April)	°C

(b) *body weight*:
typical summer weight of a British pipistrelle bat: 8gm
probable weight of same bat when going into hibernation:
gm

(c) *food consumption*:
How many insects could a colony of pipistrelle bats, numbering one hundred, consume in a night's feeding?

UNIT 5

Pollutants and Food-chains

Pollutants will, if present in sufficient quantity, have adverse effects on animals and plants. Many of these compounds do, of course, break down so rapidly that there is little risk of toxic concentrations. But compounds such as the organochlorine insecticides and polychlorinated biphenyls (PCBs) can persist long enough for living organisms to acquire considerable concentrations within their bodies. It is commonly believed that such persistent substances accumulate or concentrate as they pass along or up the food-chain. This idea of concentration developed from field observations on the occurrence of organochlorine insecticides in wildlife specimens, and most of our evidence still relates to these insecticides. There is, however, little good evidence that pollutants in general, or organochlorine insecticides in particular, do concentrate along the food-chain.

It is common knowledge that animals are selective in what they eat – one man's meat is another man's poison. The energy that animals obtain from their food is derived ultimately from the sun's radiant energy, some of which is utilized by green plants for photosynthesis. The first link in the food-chain, or food-web as it is now usually thought of, is the herbivores. These may be preyed on by carnivores, which may themselves be eaten by other predators. It is then easy to assume that persistent compounds will accumulate along the food-chain. If 10 herbivores all have, and retain, equal amounts of a persistent substance, then the carnivore that eats these 10 herbivores will contain 10 times as much of this substance as did any one of the herbivores.

But this is not a real situation, and the confusion comes possibly from the unappreciated difference between persistence in the physical environment and persistence within organisms. Organochlorine insecticides last a long time in soils – given suitable conditions 50 per cent may remain for well over 10 years. And elemental pollutants such as the heavy metals obviously persist indefinitely, either in the elemental form or in a variety of compounds. But we do not yet know of any pollutants that remain indefinitely within a living organism.

In general terms, there are two processes by which animals can remove foreign substances from their bodies: metabolism and excretion. Metabolism, or chemical change, can by itself remove some compounds from the body. It is sometimes a necessary preliminary too before a substance can be excreted. This is particularly true of compounds such as the organochlorine insecticides that are the residue left unaffected by metabolism and excretion.

Considerable studies have been made of the quantitative relationships between rate of intake and rate of loss of drugs, and we can probably use the same mathematical models for studies on the kinetics of pollutants. Such analyses have been made for organochlorine insecticides. It is appropriate to use the compartmental model – so called because the body is considered as a series of compartments, each defined as a quantity of insecticide with uniform and distinguishable kinetics of metabolism and of transport to the other compartments. In biological terms these compartments usually correspond to specific tissues or organs. The data available are mostly for mammals, but the little information we have for other groups, principally birds and fish, suggests that they are similar.

These quantitative relationships can usefully be summarized in four simple statements:

1. The concentration of insecticide in a tissue depends on the rate of intake. The higher the rate of intake, the higher the concentration in the body.
2. Eventually, if the rate of intake is constant, and continues for long enough, the concentration of insecticide in a tissue reaches a plateau level. Sometimes, months after a plateau concentration has been reached with constant rates of insecticide intake, the tissue concentrations may change again, either up or down.
3. One can deduce from the above two relationships that the concentrations in different tissues should be correlated. They are. In general terms, adipose (fat) tissue usually contains by far the highest concentrations. This is presumably related to the fat-soluble nature of the organochlorine insecticides.
4. The existence of a plateau level implies that the animal is excreting and/or metabolizing the absorbed insecticide at the same rate as it is being absorbed. In order words, the concentrations of insecticide in the various compartments are in a steady state. One would expect, therefore, that if the rate of exposure to insecticide decreases or ends, the tissue concentrations will drop. They do. Sometimes when exposure ends there is a simple exponential decline. But sometimes, in some tissues, the rate of decline is more rapid at first than later on. These results appear to be consistent with a model in which insecticide can be excreted directly from the blood, whereas insecticide in other tissues has to pass into the blood first.

[Extract from Moriarty, Frank, Principal Scientific Officer at Monks Wood Experimental Station, Faringdon, "Pollutants and Food-chains" in *New Scientist*, 16.3.72.]

EXERCISES

5.1 Skimming

Read through the passage very quickly. Do not try to understand every word; just try to get a general idea of the topic under discussion. Now choose a new title for the passage, from this list:

(a) In defence of insecticides
(b) The dangers from insecticides
(c) How organisms reject foreign substances
(d) The concentration of insecticides in animal tissue
(e) Concentration and elimination of pollutants
(f) Research into the dangers from chemicals

5.2 Understanding in detail

Choose the correct interpretation of the phrases quoted from the text.

(a) there is little risk of toxic concentration =
 (i) no real danger of toxic concentration
 (ii) some danger of toxic concentration
 (iii) a good chance of toxic concentration
(b) there is little good evidence =
 (i) some good evidence
 (ii) a small amount of good evidence
 (iii) not really any good evidence at all
(c) adipose tissue contains by far the highest concentration =
 (i) concentrations in adipose tissue are much higher than in other tissue
 (ii) concentrations in adipose tissue are nearly the highest of all tissues
 (iii) concentrations in adipose tissue are a little higher than in other tissue
(d) the little evidence we have =
 (i) some of the evidence we have
 (ii) the small amount of evidence we have
 (iii) part of the evidence we have
(e) it is common knowledge that animals are selective =
 (i) it is obvious that animals are selective
 (ii) everybody knows that animals are selective
 (iii) most people think that animals are selective
(f) one man's meat is another man's poison =
 (i) different animals like different kinds of food
 (ii) some men can eat poison
 (iii) some meat is poisonous to men

5.3 Understanding sentence structure

Fill in the blanks with the appropriate word.

After careful scrutiny, the ______ that organochlorine insecticides and ______ pollutants concentrate along the food ______ proves to be suspect. The amounts of such ______ retained by different organisms ______ depends more on differing ______ of metabolism and excretion.

5.4 Understanding sentence structure

Match an element from column A with an element from column B in order to form sentences.

	A	B
1	changes associated with age	all of their ingested insecticide
2	this experiment showed clearly	that plateau concentrations have been reached
3	analyses of various organisms	is less important than rate of elimination
4	it is an implicit assumption	are little understood
5	terrestrial predators do not retain	showed a higher concentration
6	position in the food chain	the importance of metabolism

5.5 Understanding the text: True or False?

Decide which statements are true, and which are not, according to what you have read in the text.

✗(a) All organochlorine insecticides break down rapidly.
✗(b) Animals cannot eliminate any pollutants from their bodies.
(c) Only herbivores ingest pollutants.
✓(d) Organochlorine pesticides can stay in the soil for more than 10 years.
(e) It has been established which pollutants persist indefinitely in living organisms.
✓(f) Some unwanted compounds can be eliminated by metabolism.
(g) Carnivores are regarded as the first link in the food-chain.
(h) Most of the information available about the kinetics of pollutants refers to mammals.
✓(i) There is a correlation between concentrations of insecticides in different tissues.

(j) Once a plateau level of concentration has been reached, the level can only decrease.

5.6 Making a summary

The text is divided into six paragraphs. Decide which paragraph each of the following sentences refers to, then arrange them in the right order to form a brief summary.

(a) Chemical persistence in soil is different from its persistence in organisms.
(b) Metabolism and excretion are important factors in the elimination of pollutants.
(c) Different animals eat different types of food.
(d) Research in other areas can help present studies on pollutants.
(e) There is a relationship between intake of insecticides and concentration in tissues.
(f) Chemical pollutants can become concentrated in organisms.

UNIT 6

Moths Lose Their Heads Over Synthetic Scents

Pheromones, those sexy perfumes that some insects use to lure their partners into mating, have been used more than once by wily farmers to control crop pests. The idea is to synthesize the parasite's pheromones and to spray a large quantity over an entire crop. The result is confusion – males and females fail to recognize each other and a complete generation of insects never sees the light of day.

Though it may seem a little unsophisticated, the chaos tactic does work, as the pink boll worm, which attacks cotton plants, has found to its cost in some parts of the US. The latest parasite to have its sex life threatened in this way is the moth *Eldana saccharina*, the scourge of sugar cane and maize in parts of Africa.

The moth is unusual in that it is the male that entices the female to mate rather than the other way round. On the first night of its adulthood, the male moth tries to gain the attentions of a female by secreting two chemical substances. The first is based on vanillin, the sweet-smelling chemical found in the vanilla bean and used in flavourings and perfumes. The second is called eldanolide and is a member of a group of chemicals known as lactones. These are odour-bearing substances found in certain plant products. Insects use the two pheromones to recognize the opposite sex when some distance apart.

The male embarks on his sexual advances by adopting a "calling posture". His wings begin to flutter and two abdominal hair pencils (androconia) which will later secrete the lactone pheromone unfold. During this phase of the mating process, another of the male's scent organs emits the attractant vanillin.

Aroused from afar by his provocative scents, the female starts to climb the stalks of sugar cane or maize to find the male. It is only at the moment of contact between *Eldana* male and female that the lactone aphrodisiac substance comes into play.

As soon as she identifies her partner, the female abandons her search behaviour and begins to flutter her wings. The sound thus produced is interpreted as a sign of readiness by the male who then mounts her and achieves coupling.

As a result of some work carried out at a joint laboratory of the Centre National de la Recherche Scientifique and the Institut National de la Recherche Agricole in the south of Paris, particularly in the chemical identification of eldanolide, biologists are currently at work in the Ivory Coast carrying out experiments with synthesized *Eldana* pheromones.

In addition to using pheromones for the confusion tactic, they can provide farmers with important information. Mating behaviour roughly coincides with the time the insects begin their flight. Using the synthetic eldanolide, biologists hope to trap female *Eldana*. When the number of females trapped rises rapidly, the flight is imminent.

Using this knowledge, farmers can apply insecticide at the most effective moment rather than once a week throughout the season, with the added advantage of reducing pollution as well as expense.

[Extract from "Moths lose their heads over synthetic scents" in *New Scientist*, 23 September 1982, p.830.]

EXERCISES

6.1 Skimming

Read through the passage quickly. Do not try to understand every word. Now decide what the article is about. Choose from these subjects. (More than one answer is possible.)

(a) Insects in America
(b) Pests in cotton plants
(c) The way moths breed
(d) Control of pests with insecticides
(e) The use of chemicals to eliminate damage to crops
(f) Biological pest control

6.2 Guessing

What sort of publication do you think this article first appeared in?

(a) A professional journal devoted to research into chemicals
(b) A magazine of popular science
(c) A daily newspaper
(d) The journal of a learned society devoted to biology

Based on the answer you choose, do you think the title is suitable?

Supposing the article appeared in one of the other types of publication named, choose a new title for the passage from these:

(i) Synthetic pheromones
(ii) Insect control in USA
(iii) Pests in sugar cane
(iv) Improving cotton crops
(v) Research in pest control
(vi) Love will find a way

6.3 Understanding the text

You must look at the text in detail. This will help you to understand it. Find words or phrases in the text which have similar meanings to those underlined in these sentences, and which can replace them.

(a) Clever farmers are learning new ways to control pests.
(b) Synthetic pheromones can attract insects into traps.
(c) The insect begins to agitate its wings when it recognizes a mate.
(d) Usually the female attracts the male, but with *Eldana saccharina* it is the reverse.
(e) Scientists are about to start a new programme of research.
(f) The female can detect the male from a considerable distance.

6.4 Understanding the text

Choose the right answer.

(a) The first pheromone produced by the male *Eldana saccharina* is:
  (i) androconia
  (ii) aldanolide
  (iii) vanillin
(b) Lactones are present
  (i) only in moths
  (ii) only in insects
  (iii) in some plants
(c) *Eldana saccharina* is a pest which damages
  (i) sugar cane in the USA
  (ii) maize in Africa
  (iii) cotton in France
(d) The "chaos tactic" results in
  (i) the immediate death of the insect
  (ii) the failure of males and females to mate
  (iii) an increased need for insecticides
(e) The female *Eldana saccharina* shows that she is ready to mate
  (i) by climbing the stalks of the sugar cane
  (ii) by moving her wings to produce a particular sound

(iii) by secreting an odour-bearing chemical

(f) Pheromones can be used to tell farmers

(i) how much insecticide to use

(ii) the best time to use insecticides

(iii) how to find the best insecticides

6.5 Words and phrases of similar meanings: Idioms

Three commonly-used English idioms occur in the text. Here an idiom means a group of words whose total meaning cannot be deduced from the meanings of the individual words. For example, the expression "to put someone up" meaning "to receive someone as a guest in your home" is an idiom. Look at this sentence:

When I was in New York, I didn't stay in a hotel because some friends put me up.

These idioms are not really part of a "scientific" vocabulary, but are appropriately used in this text which is written for the ordinary reader as well as for the specialist. Pick out these three idioms from the text and use one in each of the following three sentences, to replace exactly the words underlined.

(a) If a fire breaks out, people sometimes become confused and behave irrationally.

(b) When a new director is appointed, it is probable that his predecessor's latest plan is not implemented.

(c) The dog discovered too late that the meat was poisoned.

6.6 Writing a summary

The text consists of nine short paragraphs. Write one sentence for each paragraph, containing the essential information in that paragraph. Combine your sentences to form a summary of the whole passage.

UNIT 7

Weed Communities

In an intact plant community, undisturbed by human intervention, the composition of a community is mainly a function of the climate and the type of soil. Today, such original communities are very rare – they are practically limited to national parks and reservations.

Civilization has progressively transformed the conditions determining the composition of plant communities. For several thousand years vast areas of arable land have been hoed, ploughed, harrowed and grassland has been cut or grazed. During the last decades the use of chemical substances, such as fertilizers and most recently of weed killers (herbicides) has greatly influenced the composition of weed communities in farm land.

All selective herbicides have specific ranges of activity. They control the most important weeds but not all the plants of a community. The latter profit from the new free space and from the fertilizer as much as the crop does; hence they often spread rapidly and become problem weeds unless another herbicide for their eradication is found.

The soil contains enormous quantities of seeds of numerous species – up to half a million per m^2 according to scientific literature – that retain their ability to germinate for decades. Thus it may occur that weeds that were hardly noticed before emerge in masses after the elimination of their competitors. Hence, the knowledge of the composition of weed communities before selective weed killers are applied is not only of scientific interest since the plant species present in the soil in the form of seeds must be considered as potential weeds. For efficient control the identification of weeds at the seedling stage, i.e. at a time when they can still be controlled, is particularly necessary; for the choice of the appropriate herbicide depends on the composition of the weed community.

Plant sociology

Plant Sociology has become established as a separate branch of Botany during the present century. Founded originally in Switzerland, at Zurich, the school of J. Braun-Blanquet, who later worked at Montpellier in S. France, has been able to establish its concepts in this field even in competition with, for example, the Northern (Scandinavian) schools. In Germany, the work of R. Tuxen and his collaborators in particular has ensured the recognition of Plant Sociology as a branch of Botanical Science. Although the terminology and sociological-systematic aspects of the subject are continually being changed and debated, we have preferred to retain the terms established by Braun-Blanquet and the classification of weed associations made by Tuxen (1950), which are the most comprehensive of those that have so far been introduced.

In this classification, the weeds of cultivated land comprise one *class* with two *orders*, each consisting of three *alliances*. Originally, these plant-sociological units were defined using purely botanical characteristics, based on so-called characteristic species or on combinations of sociological groups, i.e. combinations of species always found together in particular situations. Today, scientists also take into account the conditions prevailing in the habitats when defining such units.

[Extract from *Documenta* (*Weed Communities*) Copyright Ciba-Geigy AG, Basle, Switzerland, 1971.]

EXERCISES

7.1 Skimming

Read through the whole passage very quickly and get an idea of what it is about.

Choose from this list the sentence which, in your opinion, best describes the main topic of the passage.

(a) Effects of chemicals on vegetation
(b) Weed control by means of herbicides
(c) The identification of weeds
(d) A study of weed communities
(e) The impact of man on plant communities
(f) The importance of studying how plants live in communities

7.2 Scanning

Now look for specific details. Go over the whole passage rapidly, but this time you are looking for particular items of information.

(a) In a square metre of soil, how many weed seeds could be present?

(b) Before the advent of Tuxen's system, on what *main basis* were weeds classified?
(c) What particular advantage is ascribed to Tuxen's classification system?
(d) Where can plant communities be found which have not been disturbed by man?
(e) At what stage of growth should weeds be controlled?
(f) How many agricultural processes are mentioned that farmland was subjected to before chemicals came into use?

7.3 Understanding the text

This exercise helps you to understand words used in the text. Here is a list of words or phrases. Find the words used in the text which have the same, or very similar meaning. For example, if the list had contained "put on", the appropriate word from the text would be "applied".

ten years
appear
at the present time
not broken
keep
area of interest
large quantities
argued about
uncommon
restricted

7.4 Understanding the text

Replace the underlined phrase with a single word. Choose the words from this list:

influenced today thus profit originally present

(a) <u>At the beginning</u> the work was done in Switzerland.
(b) Scientists <u>at the present time</u> prefer a different system.
(c) Seeds can live in the soil for years. <u>From this fact</u> we can assume that new weed problems will appear.
(d) Weeds as well as crops can <u>get an advantage</u> from fertilizer.
(e) Millions of weed seeds are <u>to be found</u> in a few square metres of soil.
(f) The use of herbicides has <u>had an effect on</u> the composition of plant communities.

7.5 True or false?

Decide which of these statements are correct, and which are not correct.

(a) It is impossible to find intact plant communities.
(b) Man had no effect on plant communities until he started using chemical substances on the soil.
(c) Selective herbicides affect only specific plants.
(d) Weeds can benefit from the application of herbicides.
(e) Plant sociology was first established in Scandinavia.
(f) Tuxen's classification system depends entirely on botanical characteristics.

7.6 Making a summary

Choose the sentences which contain the main topics of the text. Then combine them into one paragraph, using linking words from this list:

because since therefore so but as thus hence

(a) It is difficult to find intact plant communities.
(b) Originally, the composition of plant communities depended on climate and soil type.
(c) Civilization has transformed the conditions in which plants live.
(d) Farmers have worked the land for thousands of years.
(e) The use of chemicals has had a profound effect.
(f) Weed killers can be very dangerous.
(g) Selective herbicides can encourage the growth of previously unimportant weeds by eliminating their competitors.
(h) There are half a million weed seeds in a square metre of earth.
(i) Seeds can lie dormant for years.
(j) It is important to know the original composition of the weed communities.
(k) Plant sociology started in Switzerland.
(l) Plant sociology is now recognized as an important branch of Botany.
(m) It is vital to use a comprehensive classification system such as Tuxen's.
(n) Conditions of habitat as well as botanical characteristics must be considered.

UNIT 8

Genetic Manipulation

Ever since man the hunter and gatherer gave up his nomadic way of life and began to tend stock and grow crops, he has been involved with genetic manipulation. Firstly, in ignorance simply by choosing to rear particular animals or plants, which were in some way advantageous to his developing lifestyle, and then much later as the science of Genetics began to develop, man has been engaged in breeding programmes designed to produce varieties of plants and animals exhibiting the specific characteristics which fit them to his various needs.

As man's exploitation of natural resources has continued and industries have developed based on the synthetic ability of micro-organisms, particularly the bacteria and fungi, his need for knowledge of the fundamental principles of the genetics of these organisms has increased and the new science of Molecular Genetics has emerged. The discipline seeks to understand the molecular base of inheritance and the way in which the information encoded by deoxyribonucleic acid (DNA) is utilized by the living cell.

Advances in the field of recombinant DNA research over the past decade have given the geneticist the techniques required to mobilize individual genes, that is specific sequences of DNA which code the amino acid structure of single proteins and transfer them from a donor to a recipient organism, thus conferring on the recipient the ability to synthesize the gene product. This is the practice of genetic manipulation as we understand the term today and which has become a cornerstone of the new Biotechnology. Now in addition to searching in nature for wild micro-organisms capable of producing specific products, a process which is often long and tedious and sometimes unrewarding, microbial hosts can be tailored for specific purposes by introducing foreign genes into them. The source of this foreign DNA can be microbial, animal, or plant and thus microbial hosts can be converted into biosynthetic factories capable of making a wide diversity of materials needed in every aspect of our lives from food and fuel to agriculture and medicine. As well as its potential in aspects of applied biology, recombinant DNA research is an extremely powerful tool for the elucidation of gene structure and function.

Most recombinant DNA experiments are designed to transfer specific genetic information from a donor organism to a recipient cell in such a way that the newly acquired gene will be expressed and result in the production of a "foreign" protein. In order to do this the DNA to be transferred must first be isolated from the donor organism and inserted into a DNA carrier or vector molecule which will be used to transfer it into its new host.

The ease with which fragments of DNA can be cut out of large DNA molecules, present in the chromosomes of plants and animals, and inserted into vectors has been assisted greatly by the discovery within the last 20 years of a group of enzymes known as restricted endonucleases. These enzymes recognize specific base sequences on DNA molecules and cut them precisely within or near that sequence. There are currently some three hundred of these enzymes known and some forty or so are commercially available in a highly-purified form.

The enormous growth of interest and input of capital into researching the applications of recombinant DNA research over the past decade is evidence of the potential benefit to man which these techniques can provide. Independent of its use for fundamental research in molecular genetics, a field which has provided and will continue to provide invaluable information to both academic and applied geneticists, recombinant DNA technology has already made important contributions in several areas of applied science.

[Reprinted from *Biologist* (1982) 29 (4) by permission of the Institute of Biology, London.]

EXERCISES

8.1 Scanning

Read through the text and quote the exact words used by the author to define, explain or amplify the following:

(a) molecular genetics
(b) the practice of genetic manipulation
(c) individual genes
(d) a DNA carrier
(e) bacteria and fungi
(f) searching in nature for wild micro-organisms

8.2 Understanding the text

The words in column A all appear in the text. For each one, find a word in column B which is opposite in meaning.

	A		B
(a)	nomadic	(i)	useless
(b)	ignorance	(ii)	entirety
(c)	particular	(iii)	knowledge
(d)	individual	(iv)	fixed
(e)	recipient	(v)	general
(f)	donor	(vi)	group
(g)	ease	(vii)	profitable
(h)	fragment	(viii)	giver
(i)	invaluable	(ix)	difficulty
(j)	unrewarding	(x)	receiver

8.3 Finding a synonym

Find a word in the text to replace the words underlined in these sentences:

(a) This English course has been specially designed for students of biology.
(b) We must bring into use all our resources.
(c) The study of genetics is a most important part of modern biology.
(d) Technology has changed our manner of living.
(e) The increase in size of the new industry has been very rapid.
(f) Interesting work is being done in the general area of synthetic production.

8.4 Understanding the text: True or False?

Decide which of the statements are true, and which are not.

(a) The practice of genetic manipulation has a long history.
(b) The science of molecular genetics is relatively modern.
(c) Only wild micro-organisms can act as hosts to foreign DNA.
(d) Microbial hosts will not accept plant DNA.
(e) Research in recombinant DNA is valuable only to industry.
(f) In recombinant DNA experiments, DNA is transferred direct to the host from the donor.
(g) Enzymes are used to identify and isolate DNA transfer.
(h) There are 40 enzymes in the group known as restricted nucleases.
(i) A lot of time and money is being spent on research into recombinant DNA.

(j) Not much can be expected of genetic manipulation in the future.

8.5 Understanding sentence structure

Fill in the blanks.

Seven mice of no _______ to their pure-bred parents _______ been born as a _______ of a new technique _______ manipulating eggs. The technique _______ replacing the genetic material _______ a newly fertile egg _______ genetic material from another _______ . Scientists used suction to _______ the original genetic material _______ a virus to ease _______ passage of the new _______ into the egg.

8.6 Making a summary

Re-arrange the order of these sentences, so as to make a very brief summary of the original passage.

(a) The future benefits from research into recombinant DNA will be very great.
(b) The aim is to produce "foreign" protein.
(c) The new science of molecular genetics has emerged.
(d) Fragments of DNA are cut out with the aid of certain enzymes.
(e) Man has been involved with genetic manipulation for thousands of years.
(f) The hosts are converted into biosynthetic factories.
(g) It is now possible to mobilize individual genes and transfer them to a host organism.

UNIT 9

Body Temperature of Birds and Mammals

What is body temperature?

The heat produced by an animal must be transported to the surface before it can be transferred to the environment. Therefore, the surface of the organism must be at a lower temperature than the inner parts, for if the temperature were the same throughout, no heat could be transferred. The conclusion is that temperature of an organism cannot, of necessity, be uniform throughout.

If we examine where in the mammalian body heat production takes place, we find that some parts produce more heat than others. In man the organs of the chest and abdomen, although they make up less than 6 per cent of the body weight, produce 56 per cent of the total heat. If we include the brain, which in man is large and has a high heat production, we have accounted for 72 per cent or more than two-thirds, of the total heat production.

We can thus consider that the body consists of a core where most of the heat production takes place, and a much larger shell which includes skin and muscles and produces only a small fraction of the total body heat. During exercise, the situation is different, for the total metabolic rate may increase ten-fold or more. Most of this increase occurs in the muscles (including the diaphragm and other respiratory muscles).

During exercise then, for the internal temperature to remain constant, more than ten times as much heat as was produced at rest must be transported to the surface of the organism.

The inner, or core temperature remains reasonably constant, but this does not mean that the temperature throughout the core is uniform. Organs which have a high rate of heat production may be warmer than others, but they are cooled by the blood, i.e. the venous blood which leaves these organs is warmer than arterial blood. The temperature differences in the core may be as much as 0.5°C from one site to another. We therefore cannot speak about a single core temperature, but as a practical measure the deep rectal temperature is often used as a representative measure.

The surface temperature of a man who is in heat balance is always lower than the core temperature. This means that the arterial blood which flows to the shell loses heat and returns as colder venous blood. This is, of course, how most of the heat produced in the core is brought to the surface, or, in other words, how the core organs are cooled. Depending on the circumstances (external temperatures and the need for heat loss), the surface temperature varies a great deal. Also, the underlying tissues, including a large part of the muscle mass, can take on temperatures considerably below the core temperature.

What is the mean temperature of the body? We can calculate a mean body temperature from multiple measurements at various sites, provided that we have weighting procedure to apply to the various measurements. However, mean body temperature calculated from multiple measurements is physiologically relatively meaningless. We have already seen that the core temperature does not represent the heat status of the whole body, that the temperature of the shell can vary widely, and that the depth in the body to which the shell temperature extends changes drastically. A change in the temperature of the shell means that the total heat content of the body changes although the core temperature may remain constant. If a man moves from a room temperature of 35°C to 20°C, the drop in the shell temperature may involve a heat loss of 200 kcal from the shell. The magnitude of this much heat becomes clear when we realize that it corresponds roughly to 3 hours' resting metabolism of the man.

[Extract from Knut Schmidt-Nielsen, *Adaptation and Environment* (1976) published by Cambridge University Press.]

EXERCISES

9.1 Understanding the text

Find the words in the text to replace the underlined words in these sentences:

(a) The regulations are the same in all parts of the country.
(b) The experiments were invalid as the conditions were not all the same.
(c) Skin and muscle compose nearly one third of the body weight.
(d) The central part of the reactor is heavily protected.
(e) The interior of the building was destroyed by fire and only the exterior walls escaped.
(f) Air conditioning reduced the temperature in the room.

9.2 Improve your reading speed

It is important to recognize word shapes quickly, and not confuse words of similar appearance. Pick out the words which appear in the text "Body Temperature".

	1	2
(a)	weighing	weighting
(b)	vary	very
(c)	mean	meal
(d)	drastically	dramatically
(e)	physically	physiologically
(f)	necessary	necessity
(g)	underlining	underlying
(h)	sights	sites
(i)	statues	status
(j)	shell	shall

9.3 References

To understand any text, the reader must understand the way ideas are linked between sentences, and what exactly are the points being referred to. Look at this sentence:

> *The inner, or core temperature remains reasonably constant, but this does not mean that the temperature throughout the core is uniform.*

What does this refer to? It refers to the fact that the core temperature remains constant.

Find the references to the words underlined by answering the question which follows each one.

Line 2: before it can be transferred
(a) What is it?
Line 8: some parts produce more heat
(b) Some parts of what?
Line 17: the situation is different
(c) What situation?
Line 18: Most of this increase occurs in the muscles
(d) What is increased?
Line 28: the blood which leaves these organs
(e) Which organs?
Line 56: The magnitude of this much heat
(f) How much heat?

9.4 Understanding the text: True or False?

Decide which of these statements are correct, according to the text, and which are not correct.

(a) Blood in veins is always hotter than blood in arteries.
(b) The outer body of a man does not have a constant temperature.
(c) Normally the shell temperature of a man is below that of the core.
(d) Heat is removed from the internal organs by the flow of blood.
(e) It is impossible for the heat of an organism to be the same in all its parts.
(f) The brain produces 16 per cent of man's total body heat.
(g) The core temperature never varies.
(h) Less than one third of a man's body heat is produced by the shell.
(i) During exercise the core temperature of a man rises by a factor of ten.
(j) Ambient temperature has no effect on shell temperature in a man.

9.5 Using information

Here is a table giving the heat production in the major organs of a man at rest. From the information contained in the text, complete the missing data.

	Organ Weight				Heat production		
	(kg)		(% of body wt)		(cal min^{-1})	(% of total)	
Kidneys	0.290		0.45		100	7.7	
Heart	0.290		0.45		139	10.7	
Lungs	0.600		0.9		57	4.4	
Brain	1.350	5.03	2.1	7.7	208	a) □	72.4
Splanchnic organs*	2.500		3.8		435	b) □	
Skin	5.000		7.8		24	1.9	
Muscle	27.000	59.97	41.5	91.3	203	15.7	c) □
Other	27.970		42.0		130	10.0	
	65.000				1296	100.0	

*Abdominal organs, not including kidneys

9.6 Sentence structure

Fill in the blanks, to provide a description of the illustration.

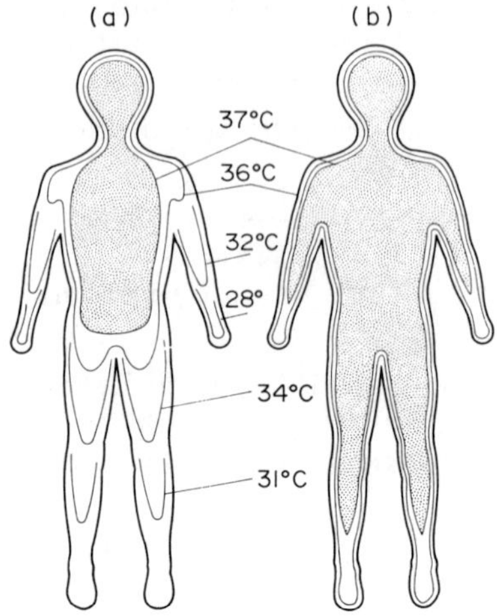

" ______ distribution in the ______ of a ______ at room ______ of ______ (left) and at ______ (______). The isotherms, which indicate ______ of equal ______ show that, at ______ room ______ , a ______ temperature of 37°C (shaded area) ______ into the ______ and arms. At 20°C ______ temperature, the ______ gradients in the ______ extend ______ the ______ and legs, and the ______ temperature is ______ to the trunk and ______ ."

UNIT 10

Developmental Plant Anatomy

Apices and apical growth

One of the characteristics most often used to distinguish plant from animal cells is the possession by the former of a cell wall. In fact a plant cell has been likened to an animal cell in a strait-jacket. In consequence plant cells very early in life tend to assume a permanent form and although in many cases they can revert to a potentially embryonic condition, normally once the semi-rigid or rigid form is attained the plant cell will no longer participate in normal growth and enlargement.

Under these circumstances the growth of plants is dependent on the continued existence of groups of cells which remain, or whose daughters remain, in an embryonic state for long periods of time. Such groups of cells (or even single cells) constitute the actively growing tissues of plants and are grouped under the general name of meristems.

The simplest meristem can be seen in a fungal filament where observations of the tip will show that growth of the hypha is the result of the prolongation of the tip cell, followed frequently by the formation of a transverse wall. An easy check on this can be made by watching the tip of a hypha of Mucor over a few hours and noting the distance of the tip from some fixed point such as a side branch.

It is a matter of some importance to realize that even in a simple filament each daughter of the apical cell may remain actively dividing for some time, but whereas the actual continuing meristematic residue will act as such, the other daughter will only contribute additional cells to the filament for a limited period of time. Meristematic activity therefore need not be, and usually is not strictly confined to the apical cell and there may be a *zone* of cell division and meristematic activity although there is only *one* apical cell (Fig. 1).

Apical differentiation

It cannot be emphasized too much that an apex is transient, and as time passes so the apex grows on and leaves behind its cells which go through extremely complex series of divisions and differentiations before assuming their final form. In fact the apex could be defined as part of the plant which destroys itself at each cell division, but from whose destruction new apical cells and the rest of the plant arise.

But the plant body is very much larger and more complex than the apex, therefore behind the apex and its dividing cells, there is a constant adjustment of form and position which must operate in such a way that the new tissues formed at the apex will exactly integrate with those already formed. Or to put it in a more definite way, the new tissues must be so formed that the whole plant will always be a continuous system with vessel accurately abutting on vessel and sieve tube meeting sieve tube.

The nicety with which these adjustments are made have called forth the obvious question of the relationship of the apex to the rest of the plant. In its simplest form, does the apex determine the shape of the plant, or is the apex induced to do things by the tissues already formed?

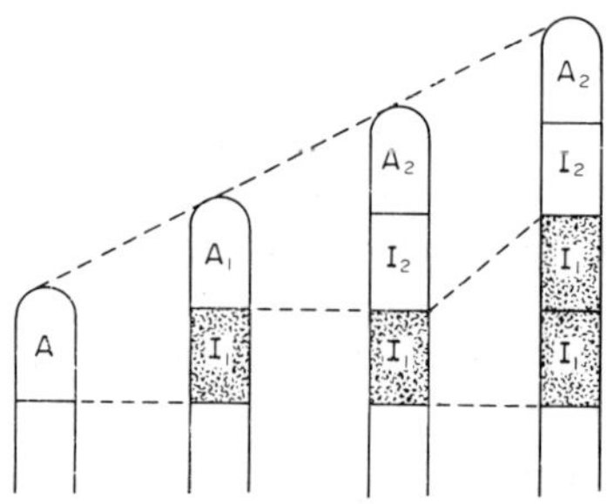

Fig. 1. *Growth of filaments.* The apical cell (A) divides to form a new apical cell A_1 and a daughter 11. A_1 may divide again to form A_2 and 1_2 and at the same time 1_1 may produce two daughter sells. This process may continue for a considerable period showing that growth and division are not limited to the apical cell itself.

[Extract from Alan R. Gemmel, *Developmental Plant Anatomy,* (*Institute of Biology's Studies in Biology,* No. 15), Edward Arnold, 1969.]

EXERCISES

10.1 Understanding the text

Find words or phrases in the text which have roughly the same meaning as the words and phrases listed below.

(a) something which stops free movement

(b) said to resemble

(c) take part in
(d) act of growing bigger
(e) extreme end
(f) not permanent

10.2 Understanding details

Here are six phrases quoted from the text. Decide which of the two explanations given is the more accurate.

(a) Line 9: under these circumstances=
  (i) in some conditions
  (ii) in the conditions just described
(b) Line 21: a matter of some importance=
  (i) something very important
  (ii) something not very important
(c) Line 29: It cannot be emphasized too much=
  (i) this fact has had too much attention paid to it
  (ii) it is very important to appreciate that (an apex is transient)
(d) Line 44: The nicety with which these adjustments are made=
  (i) the adjustments are made very accurately
  (ii) the adjustments are made very frequently
(e) Line 46: In its simplest form=
  (i) the simplest form of the apex
  (ii) the simples form of the question
(f) Line 47: is the apex induced to do things by the tissues=
  (i) does the apex do things to the tissues
  (ii) do the tissues make the apex do things

10.3 References

This exercise is of the same type as Exercise 9.3 on page 38. What do the underlined words refer to?

Line 2: the possession by <u>the former</u>
  (a) What is "the former"?
Line 3: <u>In consequence</u> plant cells . . .
  (b) In consequence of what?
Line 12: <u>Such groups</u> of cells
  (c) What sort of cells?
Line 18: An easy check on <u>this</u>
  (d) What is "this"?
Line 24: will act as <u>such</u>
  (e) Will act as what?
Line 44: <u>these</u> adjustments
  (f) What adjustments?

10.4 Sentence structure

Fill in the blanks.

Behind the actual apical ______ cells begin to change ______ assume permanent or semi-permanent ______ as they age. This ______ of differentiation must be ______ by DNA but since ______ with the same DNA ______ eventually have very differing ______ and functions one is ______ to assume that other ______ operate which influence final ______ . This is the basic ______ of differentiation about which ______ is known that is ______ .

10.5 Relationships

This exercise is of the same type as Exercise 3.6 on page 13.

This time you have to recognize the following relationships:

Cause, result, contrast, reinforcement (additional evidence introduced to support a previous statement).

	Pointer	Cause	Result	Contrast	Reinforcement
(a) Line 2	In fact				
(b) Line 3	In consequence				
(c) Line 5	although				
(d) Line 9	Under these circumstances				
(e) Line 23	whereas				
(f) Line 26	therefore				
(g) Line 28	although				
(h) Line 32	In fact				
(i) Line 34	but				

10.6 Writing a summary

Choose from among the sentences given below those which you think form the best summary of the passage and arrange them in order, so as to make a paragraph.

(a) Simple meristems can be studied in fungus.
(b) The plant body is more complex than the apex.
(c) Plant cells possess cell walls while animal cells do not.

(d) The apex destroys itself at each cell division, but gives rise to a new apical cell.
(e) Each daughter cell remains actively dividing.
(f) This forces one to question the relationship of the apex to the rest of the plant.
(g) Meristem activity is not strictly confined to the apical cell, as there is a zone of such activity.
(h) A plant cell is an animal cell in a strait-jacket.
(i) Growth depends on embryonic cells.
(j) Actively growing plant tissues are called "meristems".
(k) Behind the apex there is a constant adjustment to accommodate the new tissues.

UNIT 11

The Song of the Love-lorn Whale

"Ditties" sung by humpback whales, possibly the most complicated songs in the animal kingdom, are really "operas". And even more remarkable, these operas vary from one ocean basin to another. Over the past 4 years, a team of scientists led by H. E. Winn, travelled the Atlantic and Pacific Oceans to hear these "dialects" (*Behavioural Ecology and Sociobiology*, vol. 8, p. 41).

Early recordings made in high latitudes, where humpbacks spend the summer months feeding on tiny crustacea, indicated a rather limited vocal performance consisting of "wheezing blows" and fast "clicks". Scientists thought these were used for exploring the local terrain by echolocation.

However, later recordings made in 1971 off Bermuda during the calving season revealed the winter songs to be a more complicated, ordered sequence of sounds lasting from 6 to 30 minutes. Winn's team was surprised to see the pattern of sounds repeated over and over again.

The song is composed of a great variety of sounds – from low "elephant snores" through moans, groans, "wavery oohs" and "long yups" to chirps, cries and the "surface ratchet" which the whale "sings" immediately before surfacing. All the songs heard in one locality closely resemble one another, and up to 18 different sounds may occur in the song, linked together into short rhythmical phrases. A number of these phrases are repeated in a certain sequence or theme. The singer then progresses in an orderly fashion through several themes, before surfacing briefly for a breath of air for another rendering.

In this latest study careful comparisons were made between the spectrographs of songs recorded in the Cape Verde Islands, off Africa in the North Atlantic, and those from the West Indies, 4300 km away. Despite some minor differences the song formats were essentially the same. Similarly, in the North Pacific songs from Hawaii and Mexico, (5000 km distant) were nearly identical, though very different from the North Atlantic songs. Songs recorded off Tonga in the South Pacific (where whales migrate 6 months out off phase with those in the northern hemisphere) were unlike either of these types.

The singers are probably unpaired males, and it seems likely that the song serves some function in courtship. The hearing of whales is believed to be acute, and the songs may be heard tens or even hundreds of kilometres away, allowing the singers to advertise their suitability and desirability to potential mates and rivals over a wide area. Females may assess the suitability of future mates on the basis of a song's intricacy. Scientists have noticed that performance of some Herculean task during courtship is expected in species where a couple stays together for a long time – and the same pairs of humpbacks have been sighted year after year. The song could also be used to identify particular whales, as individual peculiarities have been detected. It remains to be seen whether the isolation of whales in different parts of the world's oceans means that genetic differences underlie the song dialects, or whether this is a purely cultural phenomenon based on learned traditions.

[Extract from *New Scientist*, 28 May, 1981 ("Monitor Section").]

EXERCISES

11.1 Skimming

Read through the text quickly. Choose a new title for the passage. Make your choice from these suggestions:

(a) Why whales sing
(b) The song of the humpback whale
(c) A study in animal communication
(d) Whales of the North Atlantic
(e) Animal radar

11.2 Understanding the text

All the words in column A occur in the text. Find words in column B which mean more or less the same thing.

	A		B
(a)	ditty	(i)	enormous
(b)	vary	(ii)	different from
(c)	calving	(iii)	song
(d)	dialect	(iv)	complexity
(e)	sequence	(v)	giving birth
(f)	fashion	(vi)	form a basis for
(g)	unlike	(vii)	single
(h)	unpaired	(viii)	differ

(i) intricacy (ix) language variant
(j) Herculean (x) series
(k) underlie (xi) manner

11.3 Understanding the text

Choose the correct answer, according to what you have read in the text.

(a) How does a single male whale probably communicate with a suitable mate?
  (i) by singing a complicated song
  (ii) by fighting with rival males
  (iii) by swimming hundreds of kilometres
(b) What is the highest number of sounds heard in one whale "song"?
  (i) thirty
  (ii) six
  (iii) eighteen
(c) How might an individual whale be identified?
  (i) by the area in which it lives
  (ii) by the song it sings
  (iii) by the distance it swims
(d) In which season of the year can North Atlantic whales be seen off Bermuda?
  (i) summer
  (ii) winter
(e) What was discovered about the songs of whales in one locality?
  (i) They all had eighteen sounds in them
  (ii) They were all very similar to each other.

11.4 Scanning

Read through the text and look for these details.

(a) What did scientists originally believe was the purpose of the whale's song?
(b) In which locality did the whales have songs similar in character to those of the West Indies?
(c) In species where a couple stays together for a long time, what seems to be expected during their courtship?
(d) How is a whale's hearing described?

11.5 Word study

The author of the text describes the sounds made by the whale in many different ways. List the words and phrases used to describe the sounds. Which particular sound does the author say is produced by the whale just before it comes up for a breath of air?

11.6 Making a summary

Pick out five key sentences and use them to make a summary of the whole passage.

UNIT 12

Island Biogeography and the Design of Natural Reserves

The flora and fauna of islands have played a central role in the development of ecological thought, from the early formulations of evolution and biogeography by Darwin and Wallace, through Mayr's demonstration of the role of geographic isolation in speciation, to the analytical theory of island biogeography pioneered by MacArthur and Wilson. Some reasons why islands have been well suited to provoking or testing theoretical ideas are that they have definite boundaries, come in many different sizes and heights and remoteness, often have relatively simple communities of plants and animals, and serve as ready-made evolutionary laboratories offering replicate "natural experiments" in community assembly.

Islands may be real islands in the ocean, or they may be virtual islands such as hilltops (where for many species the surrounding lowland presents a distributional barrier), lakes, or wooded tracts surrounded by open land. In particular, the natural reserves and wildlife refuges that are set aside from large areas bent to man's purposes may be thought of as islands in a sea of altered habitat. In view of the manifest destiny of much of the world's tropical rain forest, we may ask such questions as: How many species of Amazonian plants and animals will survive if only 1 per cent of the Amazonian rain forest can be preserved? At what rate will species be extinguished? Which species will most likely be lost? These are pressing questions, to which the theory of island biogeography holds at least some of the answers.

Preston (1962) and MacArthur and Wilson (1963, 1967) have pointed out that the number of species on an island is set by a dynamic balance between immigration and extinction. For any particular island, the net extinction rate will increase as the total number of species present (S) increases, conversely, the net rate at which new species are added – the immigration rate – will decrease as S increases.

This perception gives insight into the relation between an island's size and degree of isolation, and the equilibrium number of species on it. Species immigrate on to an island as a result of dispersal of colonists from continents or other islands: the more remote the island, the lower the immigration rate. Species established on an island run the risk of extinction due to fluctuation in population numbers: the smaller the island, the smaller is the population and the higher the extinction rate. Area also affects immigration and extinction in several other ways: through its relation to the magnitude of spatial and temporal variability in resources; by being correlated with the variety of available habitats; and by being correlated with the number of "hot spots", or sites of locally utilizable resource production for a particular species. All in all, the larger and less isolated the island, the higher is the species number at which it should equilibriate.

[Extract from Diamond, Jared, and May, "Island Biogeography and the Design of Natural Reserves" in Robert M. May, (ed.) *Theoretical Ecology – Principles and Applications*, published by Blackwell Scientific Publications, 1976.]

EXERCISES

12.1 Skimming

Read through the text very quickly and now choose the sentence which, in your opinion, best describes the subject of the passage.

(a) Islands are all threatened by destruction.
(b) The Amazonian rainforest is being gradually destroyed.
(c) Man must do something to protect wildlife.
(d) Through the study of actual islands we can learn how to provide the best wildlife reserves.
(e) Animals and plants on islands are in great danger.

12.2 Understanding the text: True or False?

From the information contained in the text, decide whether the statements are true or false.

(a) The study of island biogeography is a recent phenomenon.
(b) Islands provide good environments for evaluating theories.
(c) Only real islands surrounded by water are suitable for study.
(d) Nobody knows what is happening to the world's rainforests.
(e) Studying islands may help to preserve some endangered species.
(f) When the immigration rate rises, there is a decrease in the extinction rate in an island community.
(g) The size of an island has no effect on the extinction rate of species.

(h) There is a correlation between the immigration rate and the degree of isolation of an island.
(i) Community assembly can best be studied in a laboratory.
(j) MacArthur and Wilson were the first scientists to work on an analytical theory of island biogeography.

12.3 Improve your reading

This is an exercise in word formation. It can help you to understand unfamiliar words if you appreciate the relationship between words of different classes but with the same derivation.

For example, the verb "preserve" occurs in the text. If we add the suffix -tion we have the noun "preservation".

Here are some verbs which occur in the text. Add the suffix and form the related nouns.

extinguish	provoke
fluctuate	correlate
immigrate	

Now let us try the reverse procedure. Here are some nouns which occur in the text. What are their related verb forms?

isolation	production
perception	demonstration

Here are some more verbs. What are their related noun forms? They do not add the -tion suffix to form nouns.

survive	disperse
develop	analyse
think	equilibriate

12.4 Finding similar meanings

Replace the underlined words or phrases with a word taken from the text. In each case the replacement is a single word.

(a) It is <u>clear and obvious</u> that the rainforests are in danger of destruction.
(b) There are many <u>urgent and important</u> questions to answer.
(c) We do not know what our <u>future condition</u> will be.
(d) They chose the site because of its <u>distance and isolation</u>.
(e) We measured the <u>distance from the ground to the top</u> of the tree.

12.5 Guessing

This passage is taken from the beginning of a longer text. How do you think it continued? Choose from the following list the topics which you think were subsequently covered:

(a) the choice of suitable places for nature reserves
(b) descriptions of laboratory experiments
(c) arriving at the ideal size for a wildlife reserve
(d) correlations between species
(e) use of chemicals in agriculture
(f) finding out which species survive

12.6 Using information

This exercise is meant to help you find specific details in a text. Read the final paragraph of the passage, then answer the questions about the islands in the map drawn below.

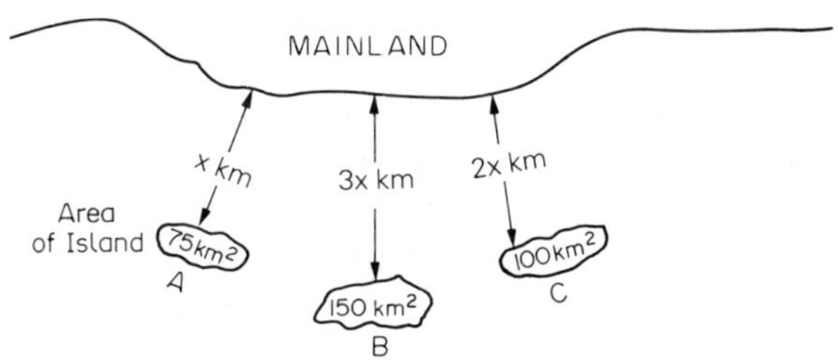

According to the information given in the text, which island should have:

(a) the highest immigration rate?
(b) the highest extinction rate?
(c) the lowest immigration rate?

UNIT 13

The Dormant State

The environment undergoes cyclic changes in which seasons favourable for growth are usually separated by periods when development must be very slow or perhaps entirely suspended, and therefore the success of a species in any habitat depends not only on its successful resistance to adverse climatic conditions, but also on an ability to synchronize its cycles of growth and reproduction with the changing seasons. It is usually essential for an organism to undergo periods of inactivity which may require the formation of special protective structures, and these periods are represented by the several states and degrees of dormancy. The development of dormancy has therefore been a major factor in evolution which has allowed the synchronization of life processes among the members of a population, and between their successive developmental stages and the seasons.

The suspension and reactivation of growth in synchrony with the seasons suggests that signals from the environment can be interpreted and used to control development and metabolism. In plants, periods of intensive growth and reproduction must occur during periods of favourable temperature and water availability. In animals, there is in addition the necessity to coincide their activities with the availability of the food species. However, an organism does not become truly dormant in immediate response to a change in the weather, but must sometimes make elaborate preparations for the dormancy state well in advance, allowing time for altered patterns of food shortage, for alterations in metabolism to produce new substances, for reproductive structures to be formed, and for developmental processes to be switched to provide protective structures. Because these processes themselves require favourable weather to continue for some time, the signal for preparation for dormancy must be interpreted in most cases before the onset of the adverse conditions themselves, and therefore may not be directly related to them. As the conditions for breaking dormancy may be experienced during the period of adversity, the factor responsible for the relief of the dormant state may be quite different from that stimulating entry into dormancy.

Response to climatic signals may result in the development of cold-hardiness and in a temporary suspension of growth, or may cause a plant to lose part, or even the whole, of its above-ground structures. Developmental pathways may be altered to produce protective bud scales in place of vegetative leaves and to cause a change from vegetative to reproductive growth. The seed habit, one of the most important evolutionary advances in the plant kingdom, requires the property of dormancy to be superimposed upon the primary role of the seed as a structure allowing the dispersal of the species. The reproductive behaviour of most living organisms has become seasonally synchronized, and, in many cases, connected with dormancy in ways which allow maximum benefit to be derived from the period of favourable weather.

Some properties of the dormant state, such as the development of cold-hardiness, tolerance to water loss and responses to similar environmental signals are to be found in a very wide range of living organisms. It is certain that dormancy as a survival mechanism must necessarily have been acquired early in evolution, and the widespread possession of one or other type of dormancy throughout the plant and animal kingdoms emphasizes the fundamental importance to life of climate and seasonal change.

[Extract from *The Dormant State*, T. A. Villiers, *Dormancy and the Survival of Plants* (Institute of Biology's *Studies in Biology 57*), Edward Arnold]

EXERCISES

13.1 Scanning for detail

Read through the text quickly, and pick out the following details.

(a) What does the author describe as "a major factor in evolution"?

(b) What environmental conditions are essential for an animal to grow and reproduce?

(c) Do the same factors both initiate and relieve dormancy?

(d) What may replace "vegetative leaves"?

(e) What two types of growth are mentioned?

(f) What value does the author give to "the seed habit"?

13.2 References

This exercise is the same type as Ex. 9.3 on p. 38. Decide what the underlined words refer to. Answer the questions.

Line 5: an ability to synchronize <u>its</u> cycles

(a) Whose cycles?

Line 9: these periods are represented . . .

(b) What are these periods?

Line 13: their successive developmental stages

(c) Whose stages?

Line 19: their activities

(d) Whose activities?

What activities?

Line 27: these processes themselves require favourable weather

(e) What are these processes?

Line 30: directly related to them

(f) What are them in this context?

13.3 Sentence structure

Fill in the blanks and complete a paragraph which appeared in the original book from which our text is taken under the heading: Defining the dormant state.

When an ordinary word is used ______ define a natural process, it often ______ loosely applied. Thus the term dormancy ______ generally used in its broadcast sense ______ describe any state of suspended activity, ______ in plants. However, although their growth ______ be suspended for a time, by ______ the temperature, actively growing plants usually ______ frozen and are killed if the ______ falls much below 0°C. The same plants, ______ prepared for winter conditions, may ______ able to resist temperatures of –80°C ______ freezing or obvious harm. It is ______ evident that true dormancy is a ______ physiological property, and the term should ______ used carefully in a scientific context.

13.4 Relationships

This is an exercise of the same type as Exercise 3.6 on page 13.

	Pointer	Logical Deduction	Contrast	Illustration	Generalization
(a) Line 3	therefore				
(b) Line 4	not only				
(c) Line 5	but also				
(d) Line 10	therefore				
(e) Line 20	However				
(f) Line 22	but				
(g) Line 30	therefore				

13.5 Using information

Complete the labels on the diagram.
Choose the appropriate words from those given below:
quiescent state growing state true dormant state dormancy environmental relative

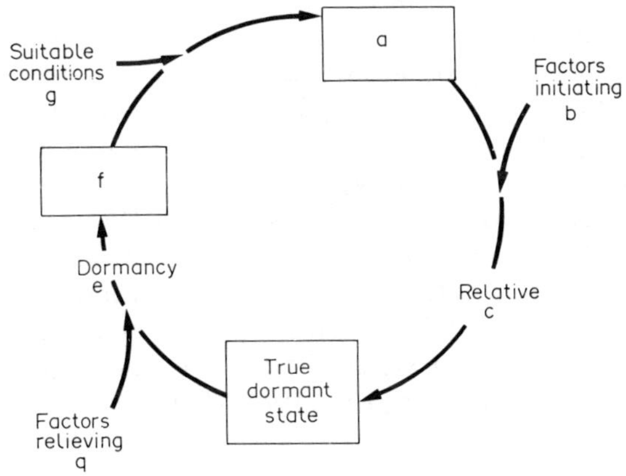

Cycle of growth and dormancy

13.6 Making a summary

Find the main topic of each paragraph and write a single sentence for each topic.

UNIT 14

Effects of Fire on Temperate Forests and Related Ecosystems: Southeastern United States

I Introduction

Studies of effects of fire on the flora and fauna of the southeastern United States have had a long and controversial history. Most of the research effort has had relevance to forest management of longleaf pine (*pinus palustris*), range management in such forests, and management of wildlifc, particularly bobwhite quail. The ecological relationships of fire and most species of plants and animals have yet to be studied in detail.

The southeastern regions of the United States (hereafter called the Southeast) ranges from the Appalachians to the sea. Its lands are some of the oldest on the continent, and yet new land is being built. It was long separated from its western plant and animal relatives and yet retained cretaceous plant "relicts". It was the refugium when glaciers reached their southern border. Subsequently man disturbed its plants and animals and turned its soil. Throughout this long period fire has been a part of its environment. Thus, in essence, many aspects of fire ecology can be studied in the Southeast for it has accumulated species from a large portion of the North American continent.

II

On a continually changing pattern, the kaleidoscope of time, space, mountains and plains, and living things, the added effects of natural fires must be superimposed. Coal beds with their attendant fossil charcoal, fusain, attest that fire was part of ancient ecosystem (Komarek, 1972). Bryson *et al* (1970) reconstructed the frontal weather systems and associated weather patterns for the late Pleistocene for the North American continent. These in many respects are quite similar to those weather systems that today create thunderstorms whose lightning discharges ignite not only forests and grasslands but buildings in farm and city as well.

Although natural fires can be started by volcanic action, spontaneous ignition, and friction, lightning appears to be the major cause of such fires. The action of this gigantic spark on forest, grassland, and field has been well documented.

III

When man became a member of the southeastern ecosystems some 20,000 years ago he became another "fire agent" in the region. Man as a fire agent was not limited to specific and short periods of ignition. He could and did set fires during long droughts and during winter months, often when lightning could not, or did not, set fires. Thus man may have set in motion other ecological changes of which we are not even aware. The litter of hardwood forests in particular is difficult to burn in the summer months, but shortly after leaf fall in autumn and after frost this fuel can easily be burned during the winter months.

IV

For present purposes fire can be considered as the combustion of recent organic matter which may vary from such compressed substances as peat to very diffuse materials such as pine "needle-drape" that envelops bushes and other understory plants. The fuel can be weathered, partly decomposed (as peat, litter, or "duff"), or recently fallen leaves, needles, grasses, and various kinds of mixtures or blends of all these and other materials.

[Extract from Chapter 8, Komarek, *Effects of Fire on Temperate Forests and Related Ecosystems: Southeastern United States*, from *Fire and Ecosystems*, ed. Kozlowski and Ahlgren, Academic Press 1974]

EXERCISES

14.1 Skimming

Choose suitable headings for Sections II, III and IV from the list given below:

Forests of the Southeast
The Appalachians
Man and Fire
Forest Management
Natural Fires
Fire and Fire Characteristics

14.2 Guessing

Based on your choice of heading from the list in 1 above, guess how Section IV continued and what topics were discussed in it. Choose from these:

(a) types of fire
(b) moisture content of fuel
(c) importance of fire precautions
(d) fire behaviour
(e) weather conditions
(f) use of fireproof materials

14.3 Sentence structure

Fill in the blanks.

Fire is the dominant ______ of forest history. The ______ majority of the forests ______ the world, excepting only ______ perpetually wet rain forests ______ the wettest belts of ______ tropics, have been burned ______ at more or less ______ intervals for many thousands ______ years.

14.4 Understanding the text

Choose the correct answer.

(a) In this text, the Southeast refers to:
  (i) a range of mountains
  (ii) a forest
  (iii) an area in the USA.

(b) Bobwhite quail are:
  (i) birds
  (ii) forest plants
  (iii) types of trees.

(c) The Southeast is a good region in which to study fire ecology because:
  (i) it has never had any fires
  (ii) it contains many plant and animal species which also live in other parts of North America
  (iii) it is quite different from the West.

(d) In the late Pleistocene age, climatic conditions in the Southeast were:
  (i) essentially the same as today
  (ii) exactly the same as today
  (iii) quite different from today.

(e) The advent of man in the Southeast resulted in:
  (i) small fires

(ii) fires during droughts
(iii) effects on the ecosystem we do not yet know about.

(f) Forest fires have as their primary fuel:
(i) coal
(ii) oil
(iii) vegetation.

14.5 References

This is a similar exercise to Exercise 9.3 on page 38.

(a) It was the refugium when glaciers reached their southern border. (line 13).
(i) plants' southern border
(ii) glaciers' southern border

(b) These in many respects are quite similar. (line 25)
(i) weather systems
(ii) effects of fire

(c) this fuel can be easily burned (line 41)
(i) wood
(ii) forest litter

(d) It was long separated from its western plant and animal relatives (line 10)
(i) the Southeast
(ii) the forest

(e) lightning appears to be the major cause of such fires (line 30)
(i) natural fires
(ii) man-made fires

(f) The action of this gigantic spark (line 31)
(i) a volcano
(ii) lightning

14.6 Making a summary

Choosing sentences, or parts of sentences, from each of the four sections, combine them to make a summary of the entire passage in under 100 words.

UNIT 15

The World of Microbes

I

The world of microbes is vast and varied. It embraces all microscopic forms, of different shapes and sizes, with the diverse colours and contrasting metabolic activities that are encountered in well-known groups of micro-organisms commonly referred to as viruses, bacteria, yeasts, algae, fungi and protozoa. Whereas some of these forms are man's best friends, others are among his worst enemies.

The friendly variety includes microbial species that ferment man's beer, destroy and transform his wastes, furnish food adjuncts to vary his diet and nutrition, and enrich the soil in which he cultivates his vegetable foods. One third of the total microbial population, on the other hand, comprises the species that attack his crops, destroy and rot his foods, cause diseases in his animals, increase his sufferings and shorten his life with loathsome and fatal scourges like tuberculosis, plague and leprosy.

Because he could not observe these friends and foes, or comprehend their interactions with him in observed phenomena, man remained ignorant of and blind to these little "animalcules" until Anton van Leeuwenhoek, a Dutch linen merchant who specialized in the grinding of powerful glass lenses, first saw and described bacteria. Since then the frontiers of the world of microbes have been gradually pushed back by the use of newer microscopes, techniques and visual reagents. Consequently, the mystery of the microbes has been revealed step by step, as the crude lens has evolved into the sophisticated electron microscope. As a result of the breakthrough into the visible world, thousands of different microbial forms were seen for the first time and man began to understand the nature of the microbes and, more particularly, their interactions with him and his environment. The simple yet spectacular contributions from the classical pioneers Pasteur and Koch helped dispel speculation, traditional lore and untested theory in the explanation of man's observed and recurrent phenomena by the establishment of scientific fact, creativity and reasoning.

II

Microbes have unwittingly been used by man in various of his arts and crafts. Recall that *Saccharomyces* were recognized and cultured as brewer's yeast in Mesopotamia by 6000 B.C., that yeasts used for making beer and bread were found in the Egyptian tombs of 2000 B.C., that the *Eber Papyrus* shows that yeast was being used in the prescriptions in the sixteenth century B.C., and that the school of Hippocrates in Cos prescribed yeasts for certain debilities. The cultural heritage of nearly every civilization is characterized by one or more fermented foods made by the souring action of microbes. *Leben* (Egypt and Syria), *dahi* (India), *yakult* (Japan), and the blue-veined cheese caused by the fungus *Penicillium roquefortii* (France) are well-known examples. Man thus deployed microbes and traditional domestic skills to make new foods, to increase their palatability, or to alter them so that they could be kept longer. The preparation of fermented foods is a widespread practice in South East Asia. Food products derived through fermentative action of microbes on seed, milk, meat, fish and vegetables are delicacies that supply protein, and that enrich predominantly starchy staple diets with vitamins and other nutrients badly needed by millions of peoples subsisting on marginal incomes. Centuries ago, the Indonesians developed market-acceptable foods with the help of microbes. Again, thousands of years ago, the Chinese developed meat-like flavours from vegetable protein through the action of *Aspergillus oryzae* on soya beans to yield the classical *shoyu* or soy sauce. Likewise, the Indonesians have their *tape*, the Japanese their *miso*, the Koreans their *kanjang*, the Vietnamese their *nuoc-mam*, the Filipinos their *patis* and the Thais their *nam-pla*.

EXERCISES

15.1 Skimming

Read through the text quickly. Now decide what the main topic is. Choose from these:

(a) The place of microbes in man's world.
(b) The use man makes of microbes.
(c) The eradication of microbes.
(d) The danger to man from microbes.

15.2 Skimming

The text is taken from a longer paper entitled The World of Microbes. Choose a suitable title for the two sections of the text from this list:

(a) Microbes and man
(b) Microbes and disease
(c) An invisible world made visible
(d) The industrial use of microbes
(e) The use of microbes over many centuries
(f) Developments in microscopes

15.3 Guessing

There are a further five sections in the original text. Try to guess what they might contain. Choose from these suggestions:

(a) How microbes can benefit the food industry
(b) Microbes can help farmers in developing countries
(c) Applications of microbiology in energy, industry and mining
(d) The most recent university studies in microbiology
(e) How to destroy microbes
(f) Benefits to public health and medicine
(g) Other useful microbes
(h) Future benefits to mankind

15.4 Understanding the text

The meaning of unfamiliar words in a text can often be deduced by noting what words they are contrasted with. For example, if you did not immediately understand the word "foe" in line 15, you could see that it contrasted with "friend", and in the first paragraph "friend" was contrasted with "enemy". Therefore, foe = enemy = something not a friend!

What are the words in the text which contrast with and explain the following?

crude lens
untested theory
spectacular
speculation
worst
mystery

15.5 Relationships

This exercise is like Exercise 3.6 on page 13.

	Pointer	Result	Logical deduction	Similarity	Generalization	Contrast	Illustration
(a) Line 5	Whereas						
(b) Line 10	on the other hand						
(c) Line 15	Because						
(d) Line 22	Conse-quently						
(e) Line 28	yet						
(f) Line 44	thus						
(g) Line 53	Again						
(h) Line 56	Likewise						

15.6 Sentence completion

Complete these sentences with words taken from the text.

(a) The number of ______ accidents has decreased, but more people have received minor injuries.
(b) Cholera was the ______ of urban society before improvements were made in public hygiene.
(c) The idea of nuclear war is ______ to all people.
(d) Lemon juice added to milk has a ______ effect.
(e) When man first used microbes, he ______ started a new industry.
(f) The ______ man regained his sight after an eye operation.

UNIT 16

Nutrients in Forests

Some aspects of nutrient cycling are more easily studied in a forest than in a water body. The more complex and massive structure of the forest makes it easier to separate from one another and measure some of the transfer rates for different means of movement and kinds of organisms. In particular, analyses of nutrient element contents of plant tissues may be combined with measurements of biomass and net production of those plant tissues. To these data should be added measurements of nutrient movement into the soil in precipitation, animal harvest of plant tissue and nutrients, and (if possible) information on nutrient release and movement in the soil.

To varying degrees the nutrients of a forest ecosystem are concentrated in the tissues of trees: the greater part of the nitrates and phosphates circulating between tree and soil may be in the trees, the greater part of the calcium and sodium in the soil. Circulation of the requisite, but limited, nutrients is relatively tight in the sense that most of the ecosystem's stock of these nutrients is held in the organisms, and the fraction in the soil is subject to rapid turnover between release (as from decaying leaves) and new uptake by roots. When a nutrient is in tight circulation in this sense: (a) the amount present in the soil at a given time is much affected by the turnover rate in the trees, (b) the nutrient may be in sufficiently short supply to affect the productivity of the trees, and (c) an addition of the nutrient to the soil as fertilizer may produce a significant, if temporary, increase in productivity.

Calcium is in loose circulation in many kinds of soils and communities. On soils derived from serpentine rock and peridotite, however, which are low in calcium, it is in tight circulation. Communities on serpentine are often stunted and reduced in structure and productivity compared with those on other adjacent soils, are of markedly different floristic composition, and are dominated by plants adapted to the low calcium levels and other special nutrient conditions on serpentine. Many species are represented on serpentine soils by ecotypes distinct from those on other soils. Phosphorus, molybdenum, and zinc are known to be limiting to community production and in tight circulation on certain soils of Australia. Tropical forests apparently have tight nutrient circulation compared with temperate forests. When the forests are cut, as in the shifting agriculture of many tropical areas, the nutrients are rapidly lost by leaching downward in the soil and erosion, and the soil loses its fertility within a few years.

The fall and decay of leaves is the most obvious means of return of nutrients to the soil. There is, however, considerably more to the nutrient return than this. In many plants some fraction of some of the nutrient elements in leaves moves back into woody twig and branch tissues before the leaves fall (while other elements may increase in the leaves up to the time of leaf fall). Plants may thus tend to conserve their supplies of certain nutrients, some of which are in short supply and tight circulation in the forest. In the long range, in the steady-state forest, the woody tissues of stems and branches must also die and decompose and release their nutrients to the soil. In the long range the total uptake from the soil and total return to the soil should be in balance.

EXERCISES

16.1(a) Guessing

What do you think was the topic under examination immediately before the passage "Nutrients in Forests"?

How do you think the passage continues? Choose a sentence, from those given below, which you think is the first sentence of the next paragraph in the original text.

(i) For two of the nutrients – potassium and sodium – a larger fraction of the element reaches the soil by washing from leaf and bark surfaces than by fall of dead leaves.
(ii) This extraction may be primarily uptake by organisms (lichens, algae, and bacteria) on bark and leaf surfaces.
(iii) The return is further affected by leaching from plant surfaces.
(iv) As in aquatic communities also, the surfaces of organisms often support films or crusts of other organisms active in nutrient uptake.
(v) Leaching refers to the effect of water dissolving materials and transporting them downward – either from plant surfaces to the soil or from upper soil levels to lower.
(vi) The greater part of this carbon was in a triple sugar, meleitose, from the secretions of aphids, a smaller part in the form of phenoilic compounds.

(vii) The water that reaches the forest floor contains also a quite significant amount of organic carbon.

(viii) Nitrogen shows a negative leaching effect, implying that it is being extracted from the water flowing across plant surfaces.

(b) Paragraph organization

Now re-arrange the remaining sentences to form the paragraph. (Two sentences, which were in the original text and gave data referring to a table, have been omitted).

16.2 Understanding the text

Find words in the text which have roughly similar meanings to these words and phrases:

(a) very large
(b) collection for use as food
(c) moving in a circle
(d) necessary
(e) decomposing
(f) of restricted growth
(g) moving from place to place
(h) very small branch
(i) over a long period
(j) cease to live

Now find words in the text which are opposite in meaning to these words (which also appear in the text).

(k) loose
(l) shifting
(m) combine
(n) return

16.3 Understanding the text: True or False?

Decide which of these statements are correct, and which are not correct, according to the information contained in the text.

(a) Size is an important factor in facilitating the study of nutrient cycles.
(b) Analysis of biomass must be excluded from any study of nutrient cycling.
(c) In the ecosystem, the greater part of the nutrients are contained within organisms.
(d) Nutrients in tight circulation are likely to remain in the soil for very long periods.
(e) Large growth can be expected from plant species on soils with little calcium.
(f) Clearing of forests has no effect on the supply of nutrients in the soil.
(g) Decomposition of leaf litter plays a large part in the nutrient cycle.
(h) Nutrient elements in leaves are not found in other tissues.

16.4 Sentence structure

Fill in the blanks.

A much larger amount ______ organic carbon falls on ______ forest floor as leaf ______ branch litter. The litter ______ attacked by soil animals, ______ , and fungi, and by ______ variously combined efforts is ______ down and degraded to ______ nutrients back into the ______ . The rate of decomposition ______ the litter and the ______ of release of the ______ nutrients varies widely, however.

16.5 Sentence structure

Rearrange these elements to make six sentences. Notice the words and phrases which occur in the passage "Nutrients in Forests" and which are here put into very different contexts.

(a) needs/prepared/to travel/efficient/a pool/the organization/at short notice/of/mobile/widely/and/technicians

(b) next/years/a/costs/to be/seems/regular/pattern/few/for/the/increase/in/the

(c) successful/profits/will not be/and/company/the/unless/the/increase/turnover/improves

(d) expanding/recent/as/some/to/theory/theory/opposed/support/tends/research/steady-state/universe/to/of/the/the/the

16.6 Using information

From information contained in the text, and in Exercise 2, label the parts of the diagram. You will need these:

decomposition uptake herb litter soil leaching precipitation branch litter ground water loss leaf litter

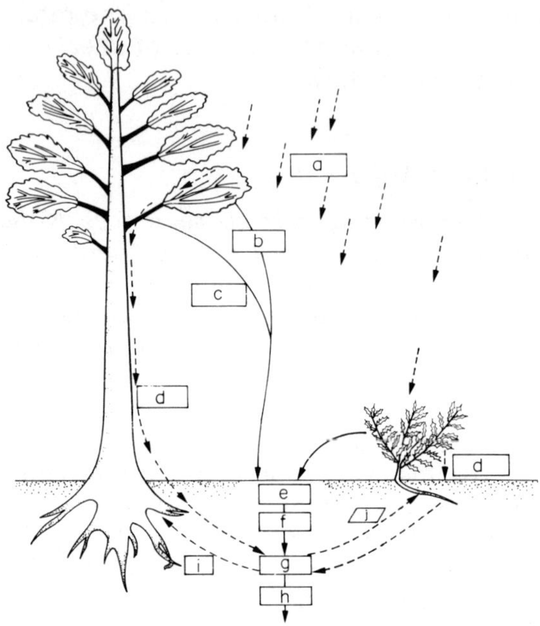

UNIT 17

Industrial Microbiology and the Advent of Genetic Engineering

Microbial cells have two main commercial applications. The first is as a source of protein, primarily for animal feed. In its commonest form this product is referred to as single-cell protein, although in fact it usually includes the entire microbial cell, the major component of which is protein.

Microbial cells are also used to carry out biological conversions, processes in which a compound is changed into a structurally related compound by means of one or more enzymes supplied by the cells. Biological conversions, also known as microbial transformations, can be accomplished with growing cells, non-growing cells, spores or even dried cells. Micro-organisms, which can carry out almost every kind of chemical reaction, have many advantages over chemical reagents. For example, many nonbiological chemical reactions call for a considerable input of energy to heat or cool the reaction vessel; in addition they are generally conducted in solvents and require inorganic catalysts, both of which may be pollutants. Finally, many nonbiological chemical reactions yield unwanted by-products that must be removed in a separate purification step.

Unlike most nonbiological chemical reactions, biological conversions proceed at biological temperatures with water as the solvent. The cells can often be immobilized on a supporting structure for continuous processing. Another valuable asset of biological conversions is their specificity: one enzyme usually catalyses only one kind of reaction at a specific site on the substrate molecule. The enzyme can also be made to select one isomer, or molecular form of a compound, in a mixture of forms to produce a single isomer of the product. These characteristics account for the high yields typical of biological conversions, which can reach 100 per cent.

The biological conversion of ethanol into a dilute solution of acetic acid (vinegar) was done in Babylon by 5000 B.C. Other important biological conversions transform isopropanol into acetone, glucose into gluconic acid and sorbitol into sorbose. (The last reaction is the only biological step in the otherwise nonbiological manufacture of ascorbic acid, vitamin C). Among the notable conversions in the pharmaceutical industry are those involved in the production of steroids. More recently, in the production of semisynthetic penicillins it became possible to replace a chemical reaction that creates pollutants with a nonpolluting biological conversion.

The most versatile large molecules manufactured by microorganisms are enzymes. These biological catalysts are important in the food and chemical industries because of their specificity, efficiency and potency under conditions of moderate temperature and acidity. Although enzymes have traditionally been extracted from plants and animals, their production by micro-organisms is increasing rapidly owing to the increasing availability of such organisms and the ease with which the yield can be improved by manipulating either the genes or the environment of the organisms. Moreover, in the microbial production of enzymes the fermentation times are short, the growth mediums are inexpensive and the screening procedures are simple.

[From *Production Methods in Industrial Microbiology* by Elmer L. Gaden, Jr., in *Industrial Microbiology and the Advent of Genetic Engineering.*]

EXERCISES

17.1 Skimming

Read through the passage quickly. The aim is to get a general idea of what it is about. Now choose from the list below what, in your opinion, the topic of the passage is:

(a) The manufacture of proteins
(b) Microbiology in the chemical industries
(c) The main uses of microbiology in industry
(d) The nature of enzymes
(e) Biological conversions
(f) Research in microorganisms

17.2 Scanning

Read the text quickly, looking for specific details. Quote the exact words *used in the text to define or explain the following:*

(a) biological conversions
(b) an isomer
(c) acetic acid
(d) enzymes

17.3 Scanning

Now find some more detailed information.

(a) What are the three industries mentioned in the text in which microbiology is said to have an important role?

(b) What are the disadvantages of nonbiological reactions when used in industrial processes?

17.4 Sentence structure

Fill in the blanks. This type of exercise helps you to appreciate sentence structure.

Recent ______ of microbially produced ______ include the use of amylases in brewing, baking ______ the ______ of textiles; ______ proteases in brewing, meat tenderizing and the manufacture ______ detergents and leather, and ______ rennin ______ cheesemaking.

A ______ recent development has been ______ combination of three microbially ______ enzymes – alpha-amylose, glucamylose ______ glucose isomerase – to obtain ______ high-fructose sweetening agent ______ cornstarch.

17.5 Relationships

This exercise is similar to Exercise 3.6 on page 13.

Time Sequence: when one event is related to another in terms of the passage of time.

	Pointer	Contrast	Reinforcement	Time sequence	Illustration
(a) Line 3	although				
(b) Line 3	in fact				
(c) Line 13	For example				
(d) Line 14	in addition				
(e) Line 16	Finally				
(f) Line 43	Although				
(g) Line 47	Moreover				

17.6 Sentence structure

Match an element from column A with an element from column B, to form sentences. Then arrange the sentences you have produced to make a short paragraph.

	A	B
(1)	the production of an enzyme can be enhanced	are replaced with starch or soybean meal
(2)	nutrients such as these	by replacing rapidly used sources of carbon and nitrogen
(3)	in the biosynthesis of enzymes it is often necessary	or a compound similar to the substrate
(4)	the manufacture of an enzyme is sometimes repressed	to exploit or bypass certain regulatory mechanisms
(5)	the inducer may be the substrate	are consumed more slowly
(6)	catabolite repression is avoided	by a natural feedback mechanism
(7)	for example, sources such as glucose and ammonia	by adding a special inducer substance

UNIT 18

Comparison of the Physical and Genetic Maps of Pseudorabies Virus Shows that the Genetic Map is Circular

Intramolecular rearrangements occur in the DNA of some herpes-viruses leading to the formation of molecules in which two regions of the genome are present in different orientations, relative to one another. Although it is clear that inverted repeated sequences present in these genomes play a role in the inversions of different segments of the genome, the exact mechanism(s) leading to this inversion has not been ascertained. Furthermore, the functional significance of the inverted repeats has also not as yet been clarified.

A comparison of the positions of mutations on the physical map with the recombinational frequencies that occur between these mutations should, in principle, shed some light on the mechanism(s) by which the viral DNA molecules come in contact with each other to exchange genetic material, as well as to clarify the role of the different regions of the genome in these interactions. These types of studies should also help in elucidating the effects the inverted, repeated sequences have on the frequency of recombination between various regions of the genome and further clarify their role in the inversion of the different segments of the genomes. Thus, a study of the genetic interactions between markers present in various parts of the genome provides a powerful tool to probe the functional organization of the complicated structure of the genomes of the herpes viruses.

Some comparisons of the physical and genetic maps of both herpes simplex virus type 1 (HSV–1) and herpes simplex virus type 2 (HSV–2) have been published. Wilkie *et al.* (1978) found little correlation between the physical and genetic maps of the HSV–1 genome; the linear order on the two maps was different. On the other hand, Chartrand *et al.* (1981) found a good correlation between the physical distance and the recombinational frequency among six mutants of HSV–2, all being located in the long unique region of the genome.

Parris *et al.* (1980) have also reported that there is a relatively good correlation between the physical and genetic maps of several mutants of HSV–1. Similar studies by Honess *et al.* (1980) yielded some evidence for the genetic circularity of the HSV–1 genetic map. However, in each of these studies, few mutants were analysed and the data need therefore to be reinforced by analyses of a larger number of mutations, both in the short and in the long sequences of the genome.

The genome of pseudorabies (Pr) virus is less complex than that of HSV (the Pr virus genome is found in only two, rather than in four, isomeric forms of HSV DNA) and thus presents distinct advantages over HSV with respect to these types of studies. For the past several years our laboratory has been involved in the isolation and characterization of mutants of Pr virus, as well as in placing these mutations on the physical and genetic maps of the genome. In this paper, we report on the results of experiments designed to determine the position of the mutations on the physical map, as well as to construct a genetic map. A comparison of the physical and genetic maps show that mutants located at the opposite ends of the physical map behave as if they were genetically closely linked, i.e. that the genetic map is circular.

[Extract from Seiji Ihara, Beth F. Ladin, and Tamar Ben-Porat, *Comparison of the Physical and Genetic Maps of Pseudorabies Virus shows that the Genetic Map is Circular*, in *Virology No. 122*, 1982. Reproduced by permission of the authors and Academic Press.]

EXERCISES

18.1 Sentence structure

These words all occur in the text:

genetic   clear   analysis   inversion   virus   mutant

Also occurring in the text are six closely related words. Use those words to complete these sentences:

(a) Further research may help to ______ the relationship between the segments.
(b) It was necessary to ______ the data before proceeding with the experiment.
(c) Scientists have identified the ______ repeat sequences.
(d) Many common diseases of man are ______ infections.
(e) A ______ is a sudden change in a gene which alters its control of development.
(f) Eye colour is determined ______.

18.2 Understanding the text: References

Decide what the underlined words refer to.

(a) relative to one another (line 3)
  (i) to the two regions
  (ii) to the molecules

(b) leading to this inversion (line 6)
  (i) the inverted repeated sequences
  (ii) the inversion of segments of the genome

(c) these interactions (line 14)
  (i) viral DNA molecules come in contact with each other
  (ii) the role of the genomes

(d) clarify their role (line 17)
  (i) the role of the inverted repeated sequences
  (ii) the role of the regions of the genome

(e) the two maps (line 27)
  (i) the physical and genetic maps of HSV–1 and HSV–2
  (ii) the physical and genetic maps of HSV–1

(f) as if they were genetically closely linked (line 49)
  (i) the maps were closely linked
  (ii) the mutants were closely linked

18.3 Relationships: Pointer words

Look at these "pointer" words and phrases. Decide whether they are introducing contrasting information or additional information in the contexts in which they are used.

(a) Although (line 4)
(b) Furthermore (line 7)
(c) as well as (line 13)
(d) also (line 15)
(e) Thus (line 18)
(f) On the other hand (line 27)
(g) also (line 31)
(h) However (line 35)
(i) therefore (line 36)
(j) rather than (line 40)

18.4 Understanding the text: True or False?

Decide which of these statements are correct, and which are not correct, according to the text.

(a) Scientists have discovered the mechanisms which result in inversion of the genome segments.
(b) It is not clear what the exact role is of the inverted repeated sequences.
(c) DNA molecules of a virus interact to exchange genetic material.
(d) Viral DNA molecules come in contact to clarify the role of the genome.

(e) Research results published in 1978 did not find any real correlation between the physical and genetic maps of the HSV–1 genomes.

(f) More mutants need to be studied before reliable data can be obtained.

18.5 Sentence structure

Find phrases in the text to complete these sentences. Each blank represents one word.

(a) At the conference, leading scientists from all over the world will ______ ______ ______ with ______ ______.

(b) The radio telescope is ______ ______ ______ which helps astronomers ______ ______ the secrets of the universe.

(c) It is hoped that their research will ______ ______ ______ on the problem and eventually lead to a solution.

(d) Recent published work has ______ ______ ______ that supports the theory of genetic circularity.

UNIT 19

Production Methods in Industrial Microbiology

In the applications of microbiology to industry the most distinctive element is usually the biological one: the exploitation of a living organism for the manufacture of a useful substance. Methods of genetic engineering promise to increase the efficiency and the versatility of the organisms on which such industries depend. It must be kept in mind, however, that a biological process can attain its full utility only when it is adapted to a context of production. Raw materials must be brought together with living cells or with components (notably enzymes) extracted from the cells; conditions that favour the biochemical transformation of the raw materials into products must be maintained; often a product must be isolated from other substances with which it is mixed. Hence industrial microbiology requires not only micro-organisms but also an environment in which the organisms can grow and a technology for handling them and their products. Both the environment and the technology are generally provided by a system of vats, pipes, pumps, valves and other devices. It follows that genetic engineering is only one factor in the success of a biological industry; the contributions of process engineering are also essential.

For any given biochemical procedure there are many ways to organize a plant of industrial scale. So far, however, only a few ways are practised; they can be divided into two broad categories: batch processes and continuous processes. In a batch process a vessel is filled with starting materials, often including the micro-organisms themselves. The biochemical conversion takes place in the vessel over a period that can range from a few hours to several days. Ultimately the vessel is emptied, the product is purified and a new batch is started. In a continuous process the raw materials are supplied and the finished products are withdrawn in a steady stream. With such a process all stages in the biochemical conversion must proceed simultaneously and at essentially the same rate. The batch process can be likened to the operation of a steel mill, whereas the continuous process stands in closer analogy to the operation of a petroleum refinery.

The choice between the batch and the continuous methods must be made on economic grounds. In general the continuous methods are best suited for a large volume of production; nevertheless, up to now most products of industrial microbiology have been made in batches. The reasons have to do in part with the biological nature of the processes and in part with the scale on which most industrial microbiology is conducted. They may continue to favour batch methods for some time to come.

[From *Production Methods in Industrial Microbiology* by Elmer L. Gaden, Jr., in *Industrial Microbiology and the Advent of Genetic Engineering.*]

EXERCISES

19.1 Skimming

(a) Pick out the key information in each paragraph.

(b) The full text had a short introduction after the main title. Which of the following do you think it was?

- (i) Biochemicals research is discovering new applications in industry. Microbiology has unlimited industrial potential.
- (ii) A whole new industry is in the making. Industrial microbiology has solved its problems of manufacture with new processes.
- (iii) Traditional practice combines with the scale on which most products are made to favour manufacture in batches. Newer, continuous methods, however, are being introduced.
- (iv) Process engineering and genetic engineering are joining forces. Industrial microbiology has a great future.
- (v) Environmental problems must be overcome before production can expand. Cost effectiveness is the dominant factor.

19.2 Scanning: Word recognition

Which word occurred in the passage?

	A	B
(a)	premise	promise
(b)	devices	divides
(c)	succeeds	success

(d) inquires requires
(e) continuous continual
(f) actor factor
(g) steal steel
(h) scream stream
(i) isolated insulated
(j) apology analogy

Now use pairs of words from columns A and B to complete these sentences.

(k) The thief used a ______ blade to open the window and ______ the jewels.
(l) His good looks were only one ______ in his success as an ______ .
(m) She gave a loud ______ when the child fell in the ______ .
(n) He lives in such an ______ part of the world that he feels ______ from all danger.
(o) The negotiations were based on the ______ that all nations would ______ to disarm.
(p) After weeks of ______ sunshine, we had two days of ______ rain.

19.3 Improve your reading

The words given below all occur in the text. Arrange them in six pairs of roughly similar meaning.

component vat environment extracted element
procedure transformation vessel withdrawn
conditions process conversion

19.4 Relationships: "Pointer" words and phrases

In previous exercises you have been given both "pointer" words and types of relationships, and asked to connect them. In this exercise you must find in the text examples of the following types of relationships and the pointer words which indicate them. Here are six types of relationship. Find an example of each one in the passage and pick out the pointer word or phrase which shows the type of relationship that is being introduced.

(a) CONCESSION: what was said before is correct, but what follows is also correct
(b) RESULT: what comes next is a result of what has been said before
(c) GENERALIZATION: a general conclusion can be drawn from information previously given

(d) CONTRAST: what comes next is in opposition to what has been said before

(e) TIME SEQUENCE: one event is related to another in terms of the passage of time

Line No.	Pointer	Relationship

19.5 References

What do the underlined words refer to?

(a) . . . on which such industries depend. (line 5)
What industries?

(b) . . . when it is adapted . . . (line 7)
When what is adapted?

(c) . . . only a few ways are practised; (line 21)
Ways of doing what?

(d) With such a process . . . (line 29)
What sort of process?

(e) They may continue . . . (line 41)
What may continue?

19.6 Sentence structure

Fill the blanks with suitable pointer words or phrases from amongst those you found in Exercise 19.4.

(a) For years the major source of polysaccharides has been plants. Recently, ________, there has been renewed interest in polysaccharides manufactured by microorganisms.

(b) The process continues until the conversion is complete. ________, the product is extracted and a new conversion is begun.

(c) The latest procedures have been very effective. ________ engineers are anxious to adopt them.

(d) Polysaccharides can now be produced microbially, ________ previously they had to be obtained from plants.

(e) The batch process is more expensive than the continuous process. ________, it is more frequently used in industry.

UNIT 20

Embryonic Development of the Hirudinid Leech *Hirudo medicinalis*: Structure, Development and Segmentation of the Germinal Plate

Most tissues and organs of clitellate annelids, including leeches, arise from the bandlets of stem cells that join to form the paired germinal bands. Right and left germinal bands coalesce on the future ventral midline of the embryo and give rise to the *germinal plate.* Histo- and organogenesis then proceed by proliferation of stem cells and their blast-cell descendants in the expanding germinal plate. The germinal plate becomes fragmented into a rostrocaudal series of distinct tissue blocks. Out of these blocks arise the metameric segmental structures known as somites and nerve ganglia, of which each is founded by a discrete number of stem cells. The germinal plate of clitellate embryos presents important structural and developmental similarities with the germ band and germinal disk of arthropod embryos.

Most of the information regarding the early stages of leech development has been obtained from embryos of the family of Glossiphonidae (order Rhychobdellidae). By contrast, very little is known about embryogenesis in the Hirudinidae (order Gnathobdellidae) despite the fact that leeches belonging to that family, especially *Hirudo medicinalis*, have been the favourite working material for the analysis of the structure and function of the relatively simple leech nervous system. This lack of developmental studies on the Hirudinidae is probably attributable to their embryos being too small for convenient experimental manipulation. Furthermore, early development of Hirudinidae is more complicated than that of the Glossiphonidae, since the low yolk content of the hirudinid egg requires the initial formation of a cryptolarva that feeds on the nutrient cocoon fluid to nourish the embryo. Since the composition of that fluid has not yet been defined, early hirudinid embryos cannot be cultured in artificial media. However, the absence of yolk presents also some experimental advantages: hirudinid embryos are sufficiently transparent to allow detailed microscopic viewing in whole mounts of the structure and arrangement of the different types of cells in the germinal plate.

A gravid *Hirudo* usually lays several cocoons during the breeding season. Each cocoon includes 5–25 eggs, about 100μm in diameter, that are bathed in the viscous cocoon fluid referred to as albumen. During the first 2 days of development of *Hirudo* at 25°C, cleavage leads to the formation of a 30– to 40–cell embryo. The majority of the cells in this embryo are small and lie at the top of three large cells, or macromeres. The macromeres no longer divide and are destined to disintegrate at later stages of development. The small cells fall into three groups. One group, lying in the anterior dorsal region of the embryo, forms the micromere cap. Cells of the micromere cap give rise to the larval mouth and envelope. A second group of cells, lying in the middle dorsal region, include the five pairs of teloblasts, which are descended from a fourth large cell. Iterated divisions of the teloblasts produce the bandlets of stem cells that give rise to the germinal plate. Four teloblast pairs are the precursors of ectoderm. One teloblast pair is the precursor of mesoderm. A third group of small cells lying in the posterior dorsal region has been reported to give rise to endoderm.

[Extract from Juan Fernandez and Gunther S. Stent, *Embryonic Development of the Hirudinid Leech "Hirudo Medicinalis": Structure, Development and Segmentation of the Germinal Plate*, in *J. Embryol. exp. Morph.*, Vol. 72, pp. 71–96.]

EXERCISES

20.1 Understanding the text

Choose the phrase which completes the sentence in accordance with the information contained in the text.

(a) The germinal plate results from:
  (i) proliferation of stem cells.
  (ii) fragmentation of tissue blocks.
  (iii) paired germinal bands.

(b) The embryogenesis of the leech has been studied mostly in:
  (i) Hirudinidae.
  (ii) Glossiphonidae.
  (iii) *Hirudo medicinalis.*

(c) The embryos of the Hirudinidae are difficult to study because they are:
  (i) small.
  (ii) complicated.

(iii) artificial.

(d) Nerve ganglia are examples of:
  (i) tissue blocks.
  (ii) germinal bands.
  (iii) metameric segmental structures.

(e) Macromeres will eventually:
  (i) divide.
  (ii) develop.
  (iii) disappear.

(f) Bandlets of stem cells arise from:
  (i) the teloblasts dividing.
  (ii) the micromeres developing.
  (iii) the small cells disappearing.

20.2 Understanding the text: True or False?

Decide which of these statements are correct, and which are not correct, according to the information contained in the text.

(a) All annelids are leeches.
(b) Somites develop from metameric segmental structures.
(c) Clitellate and arthropod embryos are alike in some ways.
(d) *H. medicinalis* has always been the favourite working material of embryologists.
(e) The eggs of the Glossiphonidae family contain more yolk than those of the Hirudinidae.
(f) Low yolk content in the hirudinid egg makes it easier to study in some ways.
(g) *H. medicinalis* has a relatively simple nervous system.
(h) The cocoon of *H. medicinalis* measures 100μm in diameter.
(i) Two days after it is laid, the egg of *H. medicinalis* has developed into a 25 cell embryo.
(j) In the embryo of *H. medicinalis* there are only three macromeres.

20.3 Relationships

What type of relationship is being indicated by these "pointer" words?

(a) By contrast (line 16)
(b) despite (line 17)
(c) Furthermore (line 23)
(d) Since (line 27)
(e) However (line 28)

20.4 Understanding the text: Words of similar meaning

These words all occur in the text. Put them into three groups, of three words in each group with a similarity of meaning.

fragmented join divisions coalesce discrete cleavage paired divide distinct

20.5 The writer's attitude

Even in a scientific text such as this one, dealing factually with a piece of biological research, the reader can detect something of the writer's attitude. Answer these three questions, and give the number(s) of the lines in the text which you think may support your answer.

(a) Do the authors seem surprised that so little research has been done on the embryo of the hirudinid leech?

(b) Do the authors think that the embryo of the hirudinid leech presents more disadvantages than advantages as an object of study?

(c) Are the authors entirely certain how endoderm arises?

ANSWER KEY

Unit 1 Respiration

1.1 (a) i (b) i (c) i
1.2 (a) breathing (b) certain types of bacteria (c) dehydrogenase enzymes (d) aerobic (e) no (f) all of it (g) three (h) more
1.3 These are TRUE: a,b,d,f,h,i,j
1.4 definition (sense), occur, confusion, converted, released, requires, within, combines, completely, rely
1.5 When flooding occurs, for example, plants can survive for several days completely immersed in water. They can survive for several weeks in waterlogged, airless soil. Anaerobic respiration is also necessary in the initial stages of germination, when the plant embryo is completely enclosed within an airtight seedcoat. The muscles of vertebrate animals can continue working for a minute or two without oxygen. This happens, for example, when an athlete runs in a race. For the first few yards his muscles respire aerobically, but soon they use up all their available oxygen.
1.6 breakdown, energy, life, supply, oxygen, complete, glucose, water, energy, respiration, oxygen, energy, glucose, substances, lactic acid, alcohol

Unit 2 DNA and the Genetic Code

2.1 c
2.2 (a) early 1950s (b) guanine, thymine (c) alternate sugar/phosphate molecules (d) how to make protein molecules from amino acids (e) enzymes (f) mitosis
2.3 These are TRUE: a,c,e,f
2.4 it, more, the, DNA, of, and, two, centre, are, one, every, opposite, is, is, rule
2.5 (1) cytosine (2) guanine (3) thymine (4) sugar (5) phosphate (6) chemical (7) G (8) T (9) C (10) C (11) A (12) C (13) G (14) T
2.6 Each gene on a DNA molecule carries instructions for making a particular type of protein. The coded instructions for protein manufacture pass from the nucleus to the cytoplasm. This task is carried out by ribonucleic acid. There are two types of this substance. They are called messenger RNA and transfer RNA.

Unit 3 The General Sensory System

3.1 (a) II (b) ii (c) the title of the original *Introduction to Biology*
3.2 (a) higher (b) touch organs (c) an electrical impulse (d) pressure, heat, cold (e) front of the upper arm (f) the lobe
3.3 b,a,e,d,c This paragraph followed the given reading passage
3.4 These were contrasted: outside/inside, simpler/higher, darkness/light, heat/cold
These were compared: burned/pricked, automatic/reflex, eliminated/suppressed
3.5 These are TRUE: c,d,e
3.6 (a) contrast (b) illustration (c) logical deduction (d) generalization (e) result (f) illustration (g) contrast (h) contrast

Unit 4 Focus on Bats

4.1 c
4.2 (a) suckled (b) unlike (c) weaned (d) nursery (e) pests (f) hibernating (g) roost (h) hollow
4.3 These are TRUE: c,e,f,h
4.4 The answer is YES to the following: b,c,d,g
4.5 pests, are, persistent, make, if/whether, bad, which, man, something, with, desirable
4.6 (a) 42°C, 32°C, 18°C, 0°C (b) 10.3gm (c) 350,000

Unit 5 Pollutants and Food-chains

5.1 e *or* d
5.2 (a) i (b) iii (c) i (d) ii (e) ii (f) i
5.3 belief, other, chain, chemicals, probably, rates
5.4 Al + B4, A2 + B6, A3 + B5, A4 + B2, A5 + B1, A6 + B3
5.5 These are TRUE: d,f,h,i
5.6 f,c,a,b,d,e

Unit 6 Moths Lose Their Heads Over Synthetic Scents

6.1 e *or* f
6.2 b Titles: a/i, c/vi, d/v
6.3 (a) wily (b) lure (c) flutter (d) the other way round (e) embark on (f) afar
6.4 (a) iii (b) iii (c) ii (d) ii (e) ii (f) ii
6.5 (a) lose their heads (b) never sees the light of day (c) found to its cost
6.6 Suggested summary: Synthetic pheromones can be used for pest control by confusing the sexes. It has been tried in the USA. The male moth *Eldana saccharina* excretes chemical substances and begins his sexual advances. The female indicates her readiness and mating is achieved. Scientists in France and Africa are working on synthetic pheromones. These substances are useful to farmers who use them to attract insects and thus determine the best time to apply insecticides.

Unit 7 Weed Communities

7.1 f
7.2 (a) up to half a million per m^2 (b) on purely botanical characteristics (c) it is the most comprehensive so far (d) in national parks and reserves (e) the seedling stage (f) Five: hoeing, ploughing, harrowing, cutting, grazing
7.3 decades, emerge, today, intact, retain, field, masses, debated, rare, limited
7.4 (a) originally (b) today (c) thus (d) profit (e) present (f) influenced
7.5 These are TRUE: c,d
7.6 b + but + c. e + since + g. Because + i + j.
1. n + therefore + m.

Unit 8 Genetic Manipulation

8.1 (a) the discipline seeks to understand the molecular base of inheritance (b) the ability to synthesize the gene product (c) specific sequences of DNA (d) vector molecule (e) micro-organisms (f) a process which is often long and tedious and sometimes unrewarding

8.2 a–iv, b–iii, c–v, d–vi, e–viii, f–x, g–ix, h–ii, i–i, j–vii

8.3 (a) tailored (b) mobilize (c) cornerstone (d) lifestyle (e) growth (f) field

8.4 The following are TRUE: a,b,g,i

8.5 resemblance, have, result, of, involves, in, with, egg, remove, and, the, genes

8.6 e,c,g,f,b,d,a

Unit 9 Body Temperature of Birds and Mammals

9.1 (a) throughout (b) uniform (c) make up (d) core (e) shell (f) cooled

9.2 These words appear in the text: 2a, 1b, 1c, 1d, 2e, 2f, 2g, 2h, 2i, 1j

9.3 (a) heat (b) parts of the mammalian body (c) where heat production takes place (d) total metabolic rate (e) organs which have a high rate of heat production (f) 200 kcal

9.4 These are TRUE: b,c,d,e,f,h

9.5 (a) 16 (b) 33.6 (c) 27.6

9.6 Temperature, body, man, temperatures, 20°C, 35°C, right, sites, temperature, 35°C, temperature, core, extends, legs, room, temperature, shell, throughout, arms, core, restricted, head

Unit 10 Developmental Plant Anatomy

10.1 (a) strait-jacket (b) likened to (c) participate in (d) enlargement (e) tip (f) transient

10.2 a–ii, b–i, c–ii, d–i, e–ii, f–ii

10.3 (a) plant cells (b) inconsequence of having a cell wall (c) cells which remain in an embryonic state (d) prolongation of the tip cell followed by formation of a transverse wall (e) as an apical cell (f) new tissues formed so that the whole plant is a continuous system

10.4 region, and, form, process, mediated, cells, may, structures, forced, factors, form, problem, little, positive

10.5 (a) reinforcement (b) result (c) contrast (d) cause (e) contrast (f) result (g) contrast (h) reinforcement (i) contrast

10.6 c,j,a,g,d,k,f

Unit 11 The Song of the Love-lorn Whale

11.1 a,b

11.2 a–iii, b–viii, c–v, d–ix, e–x, f–xi, g–ii, h–vii, i–iv, j–i, k–vi,

11.3 a–i, b–iii, c–ii, d–ii, e–ii

11.4 (a) echolocation (b) Cape Verde Islands off Africa (c) the performance of a Herculean task (something very big and difficult) (d) it is believed to be acute (very good)

11.5 wheezing blows, clicks, elephant snores, moans, groans, wavery oohs, long yups, chirps, cries, (and just before it comes to the surface for air) the surface ratchet

11.6 Suggested key sentences: (Whale) "operas" vary from one ocean basin to another. The song is composed of a great variety of sounds. All the songs heard in one locality closely resemble one another. It seems likely that the song serves some function in courtship. It remains to be seen . . . whether . . . genetic differences underlie the song dialects, or whether this is a purely cultural phenomenon.

Unit 12 Island Biogeography and the Design of Natural Reserves

12.1 d
12.2 These are TRUE: b,e,h,j
12.3 extinction, fluctuation, immigration, provocation, correlation
isolate, perceive, produce, demonstrate
survival, development, thought, dispersal, analysis, equilibrium
12.4 (a) manifest (b) pressing (c) destiny
(d) remoteness (e) height
12.5 a,c,d
12.6 a–A, b–A, c–B

Unit 13 The Dormant State

13.1 (a) the development of dormancy (b) favourable temperature, water availability, availability of the food species (c) no (d) protective bud scales (e) vegetative, reproductive (f) one of the most important evolutionary advances
13.2 (a) a species' cycles (b) periods of inactivity (c) members of a population (d) animals' activities; intense growth and reproduction (e) alterations in metabolism, formation of reproductive structures, developmental processes (f) the processes
13.3 to, becomes, is, to, especially, can, lowering, become, temperature, when, be, without, therefore, special, be
13.4 (a) logical deduction (b) illustration (c) illustration (d) logical deduction (e) contrast (f) contrast (g) generalization
13.5 (a) growing state (b) dormancy (c) dormancy (d) dormancy (e) relative (f) quiescent state (g) environmental
13.6 Suggested summary: Organisms must synchronize their life cycle with the seasons. Suspension and reactivation of growth appear to respond to environmental signals. Climatic signals probably affect development. Dormancy shows the importance of climate and seasonal change.

Unit 14 Effects of Fire on Temperate Forests and Related Ecosystems: Southeastern United States

14.1 II Natural Fires III Man and Fire IV Fire and Fire Characteristics
14.2 a,b,d,e
14.3 fact, great, of, the, and, the, over, frequent, of
14.4 a–iii, b–i, c–ii, d–i, e–iii, f–iii
14.5 a–ii, b–i, c–ii, d–ii, e–i, f–ii
14.6 Suggested summary: The ecological relationships of fire and most species of plants and animals have yet to be studied in detail. Many aspects of fire ecology can be studied in the Southeast. Fire was part of the ancient ecosystem and weather systems and patterns for the late Pleistocene are similar to those today that start forest fires. When man became a member of the southeastern ecosystem he became another "fire agent", not limited to specific and short periods. Fire can be considered as the combustion of recent organic matter.

Unit 15 The World of Microbes

15.1 b,a
15.2 c,e
15.3 b,c,f,g,h
15.4 electron microscope, scientific fact, simple, reasoning, best, revealed

15.5 (a) contrast (b) contrast (c) result (d) logical deduction (e) contrast (f) logical deduction (g) illustration (h) similarity
15.6 (a) fatal (b) scourge (c) loathsome (d) souring (e) unwittingly (f) blind

Unit 16 Nutrients in Forests

16.1 (a) nutrient cycling in water body (b) iii, v, i, viii, ii, iv, vii, vi
16.2 (a) massive (b) harvest (c) cycling (d) requisite (e) decaying (f) stunted (g) shifting (h) twig (i) in the long range (j) die (k) tight (l) steady state (m) separate (n) uptake
16.3 These are TRUE: a,g
16.4 of, the, and, is, bacteria, their, broken, release, soil, of, rates, different
16.5 (a) The organization needs a pool of mobile and efficient technicians prepared to travel widely
(b) A regular increase in costs seems to be the pattern for the next few years.
(c) The company will not be successful unless the turnover improves and profits increase.
(d) Some recent research tends to support the steady-state theory as opposed to the theory of the expanding universe.
16.6 (a) precipitation (b) leaf litter (c) branch litter (d) leaching (e) herb litter (f) decomposition (g) soil (h) ground water loss (i) uptake

Unit 17 Industrial Microbiology and the Advent of Genetic Engineering

17.1 e,c
17.2 (a) microbial transformations (b) molecular form of a compound (c) vinegar (d) biological catalysts
17.3 (a) pharmaceutical, food and chemical industries (b) considerable input of energy, catalysts may be pollutants, unwanted by-products
17.4 applications, enzymes, and, manufacture, of, of, of, in, major, the, produced, and, a, from
17.5 (a) contrast (b) reinforcement (c) illustration (d) reinforcement (e) time sequence (f) contrast (g) reinforcement
17.6 A3–B4, A1–B7, A5–B3, A4–B6, A6–B2, A7–B1, A2–B5

Unit 18 Comparison of the Physical and Genetic Maps of Pseudorabies Virus Shows that the Genetic Map is Circular

18.1 (a) clarify (b) analyze (c) inverted (d) viral (e) mutation (f) genetically
18.2 a–i, b–ii, c–i, d–i, e–ii, f–ii
18.3 These introduce contrasting information: a,f,h,j
18.4 These are TRUE: b,c,e,f
18.5 (a) come in contact/one another (b) a powerful tool/to probe (c) shed some light (d) yielded some evidence

Unit 19 Production Methods in Industrial Microbiology

19.1 (a) (Suggested answer) Industrial microbiology requires micro-organisms, an environment in which they can grow and a technology for handling them. Bio-

chemical plants are broadly organized for either batch process or continuous process. The choice of process depends on cost and scale of production.
(b) iii

19.2 These appeared in the text: a–B, b–A, c–B, d–B, e–A, f–B, g–B, h–B, i–A, j–B
(k) steel/steal (l) factor/actor (m) scream/stream (n) isolated/insulated
(o) premise/promise (p) continual/continuous

19.3 component/element vat/vessel environment/conditions extracted/withdrawn procedure/process transformation/conversion

19.4 (a) however (line 6), nevertheless (line 37) (b) It follows that (line 17)
(c) Hence (line 12) (d) whereas (line 32) (e) Ultimately (line 26)

19.5 (a) industries using microbiology in the manufacture of a useful substance
(b) a biological process
(c) organizing a plant of industrial scale for a biochemical procedure
(d) a continuous process
(e) the reasons for choosing one process rather than the other

19.6 (a) however (b) Ultimately (c) Hence (d) whereas (e) However

Unit 20 Embryonic Development of the Hirudinid Leech *Hirudo medicinalis*: Structure, Development and Segmentation of the Germinal Plate

20.1 a–i, b–ii, c–i, d–iii, e–iii, f–i

20.2 These are TRUE: c,e,f,g

20.3 (a) contrast (b) concession (c) reinforcement (d) logical deduction
(e) contrast

20.4 fragmented/discrete/distinct join/coalesce/paired divide/divisions/cleavage

20.5 (a) yes (lines 16–19) (b) no (lines 27–31) (c) no (last sentence of text authors say: " . . . have been reported to give rise to . . . " not: " . . . (they) give rise to . . . ")